PREVENTION OF BUG BITES, STINGS, AND DISEASE

DANIEL STRICKMAN

STEPHEN P. FRANCES

MUSTAPHA DEBBOUN

with illustrations by Rachel Strickman

PREVENTION OF BUG BITES, STINGS, AND DISEASE

OXFORD
UNIVERSITY PRESS
2009

OXFORD
UNIVERSITY PRESS

Oxford University Press, Inc., publishes works that further
Oxford University's objective of excellence
in research, scholarship, and education.

Oxford New York
Auckland Cape Town Dar es Salaam Hong Kong Karachi
Kuala Lumpur Madrid Melbourne Mexico City Nairobi
New Delhi Shanghai Taipei Toronto

With offices in
Argentina Austria Brazil Chile Czech Republic France Greece
Guatemala Hungary Italy Japan Poland Portugal Singapore
South Korea Switzerland Thailand Turkey Ukraine Vietnam

Published by Oxford University Press, Inc.
198 Madison Avenue, New York, New York 10016

www.oup.com

Library of Congress Cataloging-in-Publication Data
Strickman, Daniel.
Prevention of bug bites, stings, and disease / Daniel Strickman, Stephen P.
Frances, Mustapha Debboun ; with illustrations by Rachel Strickman.
p. cm.
Includes bibliographical references and index.
ISBN 978-0-19-536577-1; 978-0-19-536578-8 (pbk.)
1. Arthropoda. 2. Bites and stings. I. Frances, Stephen P.
II. Debboun, Mustapha. III. Title.
QL434.S95 2009
613.6–dc22 2008022390

This book is dedicated with love to my parents, Alfred and Lorraine, who tolerated toads, pelicans, and many jars, bottles, and boxes of bugs—the early foundation of an entomologist.

—Daniel Strickman

This book is dedicated with love to my wife, Alicia.

—Stephen Frances

This book is dedicated with love to my four brothers and four sisters; my beautiful and loving wife, Natalie; our wonderful children, Ameena, Adam, and David; and to the memory of my parents.

—Mustapha Debboun

CONTENTS

FOREWORD

Humans apparently lack the innate ability to recognize creatures which might harm us, unless they are obviously menacing predators such as tigers and sharks. Little bugs that want to feed on us or to injure us for their self-defense are mostly inconspicuous and stealthy in their approaches to our bodies. We learn about them by the painful experiences of their bites and stings, despite parental warnings. Children have a natural curiosity to touch and play with almost everything, until they understand what to fear and avoid. And most adults remain only vaguely aware of how to recognize hazardous bugs and to distinguish them from the majority of harmless species. For the majority of folks, therefore, this helpful book explains what to watch out for and how to limit the risks of regrettable encounters with bugs that are capable of hurting us.

Whereas the aesthetics of beautiful butterflies and useful ladybugs are widely appreciated, some people become generally phobic of "creepy-crawlies" and may even experience *delusory parasitosis*[1] in which they imagine themselves to be infested with microscopic bugs, wrongly blaming specks of dirt, fibers, and scabs formed where they scratch habitually.[2] Hopefully this book offers particular comfort for entomophobic and other individuals seeking guidance on how to avoid, repel, and identify harmful bugs. Culturally, the fear of bugs has been exploited in numerous fictional books and films: almost 100 scary movies[3] featuring giant bugs exaggerate and distort the realities described in this volume.

Few potentially harmful bugs actually appear to be dangerous. Aggressive wasps (yellow jackets) and bees have conspicuous patterns of warning coloration coupled with venomous stings, deterring us from interfering with their nests. Yet during millennia of beekeeping, we have not managed to produce stingless honey bees. Humanity's general aversion to insects grows from the accumulated knowledge that some are directly harmful (as described in this book) or from our fear of the diseases they transmit. Western civilization discourages us from eating tasty nutritious insects[4,5] or their products, except for honey. The 2007 *Bee Movie*[6] gives a lighthearted account of why bees readily fight back; indeed, the more irritable Africanized honey bees spreading across the Americas have been justifiably called killer bees[7]—giving us pause for thought about the problem of colony collapse disorder.[8] Unfortunately, there is no easy remedy for bee and wasp stings.

The authors of this book are well suited to provide us with defensive information about bugs, for they are all career military officers: retired U.S. Army Colonel Daniel Strickman, grandson of immigrants, now leads the U.S. Department of Agriculture's Agricultural Research Service's national Program for Veterinary, Medical, and Urban Entomology,[9] which is based in Maryland; Major Stephen Frances serves as head of Vector Surveillance and Control[10] for the Australian Army Malaria Institute, Brisbane, Queensland; while Colonel Mustapha Debboun, born in Tangier, Morocco, serves as chief of the Medical Zoology Branch in the Department of Preventive Health Services of the Academy of Health Sciences at the U.S. Army Medical Department Center and School in Fort Sam Houston, Texas.[11] They have devoted their professional lives to battling with bugs, and their textbook, *Insect Repellents: Principles, Methods, and Uses*,[12] became an immediate classic. Hence we can be confident that this more popular style of book, amply illustrated by Rachel Strickman, will provide the public with all the information required to raise awareness about our bug enemies and how to protect ourselves against them anywhere in the world.

—Graham B. White

REFERENCES

1. Nancy C. Hinkle (2000). Delusory parasitosis. *American Entomologist* 46(1): 17–25.

2. Claire Panosian Dunavan (2006). Vital signs: Bugs are crawling in my skin. *Discover Magazine* 27(12): 26–27.
3. May Berenbaum and Richard J. Leskosky (2003). Insects in movies. In V. H. Resh and R.T. Cardé, eds., *Encyclopedia of Insects*. Academic: New York, 756–762.
4. Brian Morris (2008). Insects as food among hunter-gatherers. *Anthropology Today* 24(1): 6–8.
5. Vane-Wright, R. I. (1991). Why not eat insects? *Bulletin of Entomological Research* 81: 1–4.
6. *Bee Movie* (2007). Dreamworks Animation, Paramount distribution. www.beemovie.com.
7. See www.invasivespeciesinfo.gov/animals/afrhonbee.shtml.
8. Questions and answers: Colony collapse disorder. www.ars.usda.gov/News/docs.htm?docid=15572.
9. See www.ars.usda.gov/research/programs/programs.htm?NP_CODE=104.
10. For example, see Stephen P. Frances and Robert D. Cooper (2007). Personal protective measures against mosquitoes: Insecticide-treated uniforms, bednets and tents. *ADF Health* 8(11): 50–56.
11. See www.cs.amedd.army.mil/details.aspx?dt=175.
12. Mustapha Debboun, Stephen P. Frances, and Daniel Strickman (2007). *Insect Repellents: Principles, Methods, and Uses*. CRC Press: Boca Raton, FL.

PREFACE

Mosquitoes, ticks, horse flies, scorpions, wasps, and many related creatures are fascinating organisms that make part of their livings by taking blood or by using venom to subdue prey and defend themselves. Most humans do not want to be the object of these bugs' attentions because the pain and aggravation caused by bites and stings distract people from their enjoyment or work. More seriously, fatal infections and dangerous venoms injected by bugs in some regions are strong motivations to avoid as many bites as possible. This book gives anyone, anywhere in the world, the information needed to do the best job possible of preventing bug bites and stings. Somewhere along the way, we hope that many readers will develop a healthy interest in insects, mites, ticks, spiders, and scorpions. The material in this book was extracted from solid science and worldwide professional experience and has been packaged so that any curious or concerned person can easily access these complicated subjects. You won't need to know any special vocabulary to understand *Prevention of Bug Bites, Stings, and Disease*, and the illustrations will emphasize the most significant points.

In order to avoid bites and stings, you need to understand how and why you get them. Insects, mites, ticks, and other members of the joint-legged animals known as *arthropods* are the bugs we want to understand. There are many types, and each type has its own habits, its own favorite places, and its own manner of interacting with people. You really must know what kinds of bugs are likely to bite or sting in order to do the best job of making the problem go away.

Cosmologists study the universe, geologists study the earth, biologists study life, botanists study plants, and zoologists study animals. Scientists get even more specialized when they study a single group of animals; *entomology* is the study of arthropods and *entomologists* are the folks who study bugs. Although entomology is a specialized field, it concerns about a million described and probably 9 million undescribed species—that's over 80% of all life on earth!

Entomologists have been busy for hundreds of years keeping up with this impressive diversity, inventing terminology for the kinds of bugs, what they look like, and how they live. Insects were a common target for the father of modern taxonomy, Linnaeus, a Swedish scientist who invented our current system of naming species with a Latin or Greek name divided into a general category (the genus) and a specific type (the species). We might blame Linnaeus for the intellectually demanding task of memorizing unfamiliar names like *Aedes aegypti* or *Leiurus quinquestriatus*, but this kind of naming has served biologists for over two centuries as the bookkeeping system for the diversity of life. Fortunately for the 99.9% of humanity who do not spend their lives studying entomology, common language provides a pretty good list of arthropod names, especially the relatively few species that bite or sting. *Prevention of Bug Bites, Stings, and Disease* provides an illustrated, fact-based guide to identification; any reader should be able to identify the bug that bites (maybe before it bites!) anywhere in the world without the need to remember a single scientific term. Line drawings and representative color photographs will help anyone put a bug in its place. The book also provides advice on how to get help from a professional entomologist and how to identify the cause of a bite when the bug is long gone.

Knowing the likely location of biting bugs is a good first step either to avoid them or to be prepared to deal with them. Armed with better knowledge of the kinds of biting arthropods, you should be able to avoid some of them simply by not being there. A worldwide tour of the hot spots for bites and the specific kinds of places where bugs are found will let you know what to avoid and how to be prepared. Most bites are no more than a slight annoyance, but a tiny fraction of them can cause serious damage in the form of disease or poisoning.

We will review the importance of different types of bites so that people know how much energy (and money) they should apply to each problem.

As much as people might hope for it, there is no single, effective action to prevent bug bites and stings. A good first step is to stop them at their sources in your own home. We start at the top of the house, go all the way to the bottom, and look around the outside for every possible source of noxious arthropods that might be fixed, eliminated, or made bug-unfriendly. Although some people are uncomfortable with insecticides, a little poison in the right places can do wonders to protect you from bugs. After reading about the kinds of insecticides available and how to use them safely, you should be able to use insecticides with confidence, whether in your home or while traveling.

Elimination of the bugs, however, is not always the best answer. Sometimes elimination is impossible, sometimes it is not practical, sometimes it is too expensive, and sometimes it causes more harm to a beneficial insect than good for the person. Erecting barriers between our tender bodies and the biting arthropods can really help. Those barriers might be the use of screens on windows, sleeping under a bed net, burning smoky coils, or wearing proper clothing. Whatever the barrier, some things work and some things don't. The chapters on barriers and clothing take the mystery out of matching the right method with the right bug.

Any trip to a drugstore, grocery, or outdoors shop will turn up rows of brightly colored bottles and tubes claiming to repel a wide variety of biting arthropods for some number of hours. Repellents generally work well, but there are many choices to be made based on individual preferences, the type of bug likely to be doing the biting, and the activity of the person involved. We will give a clear picture of this complicated subject using the best science available. Natural repellents made from plant extracts appeal to many people because they usually have a nice aroma and seem less harsh, but only a few of the natural active ingredients even come close to the effectiveness of synthetic repellent products, such as DEET, which dominates the market. There are other highly effective repellents that offer some advantages over DEET. What is more, the formulation of the repellent makes a big difference to its effectiveness, ease of use, and general pleasantness (or unpleasantness) when applied to the skin. We will review the repellent products available around the world so that you can use the right one in the right place for the right bug—and know the safety of the repellent application.

From the bugs' standpoint, bites and stings are sometimes about defense rather than feeding. Venomous creatures like scorpions, spiders, centipedes, urticating caterpillars, wasps, ants, and bees can also cause a lot of trouble for people. Some toxic bites and stings are painful, others cause long-lasting damage, and some can even kill. These creatures normally feed on other insects, making them a useful and necessary part of the earth's ecology; therefore, complete elimination is seldom a reasonable goal. We can avoid their bites and stings by understanding which of them are actually dangerous and learning about their habits and habitats. Avoidance and barriers are usually enough to prevent most bites and stings, but occasionally it is necessary to kill the bugs with pesticides.

Will bug protection ever get better? This volume tries to make it better by getting the right information to the right people. Scientists are working on completely new products that could make a big difference to ordinary people. We will give a glimpse into a better future based on recent discoveries and a big dose of speculation.

The lessons of *Prevention of Bug Bites, Stings, and Disease* are brought together in practical examples seen through the eyes of Melissa and George Carter in chapter 15. Melissa is a hypothetical grandmother who overcomes bug after bug as she and her family fix their home, travel to exotic destinations, and go camping. Her soldier son accepts his mother's advice (this is a fictional example!) to protect himself against everything from sand fleas to ticks. Protection from bites involves a lot of knowledge, but anyone can do it.

So, take a few hours to read this book and explore the complex world of biting arthropods, the damage they cause, and the many options we have for their control. Use those lessons, enjoy the world, and don't let the bugs bite!

ACKNOWLEDGMENTS

We would like to thank our many colleagues who have generously shared their knowledge and experiences with us throughout our careers. No one can be everywhere at once, though it is remarkable how the net of professional contacts can extend knowledge beyond the individual. Sometimes it is difficult to know who taught what to whom, but it is the rare fact in this book that has been presented without the help of others. In particular, Daniel Strickman would like to give credit to some important teachers in his life as an entomologist: the late Dr. William Horsfall, his graduate advisor; Dr. Ronald Rosenberg, a good friend and demanding boss; Colonel Albert Kinkead, a heroic example of an effective soldier; Linda Strickman, his wife and an insightful biologist; and Rachel Strickman, his daughter and an inventive illustrator. Stephen Frances would like to thank Professor Dennis Shanks, director, and Dr. Bob Cooper, commanding officer, both of the Australian Army's Malaria Institute, for their support over a number of years. Mustapha Debboun would like to thank Dr. Jerome Klun, Colonel Eugene Gerberg, Major General George Weightman, Brigadier General William Bester, Colonel Terry Klein, Colonel Stephen Berté, Colonel Randy Perry, Colonel John Ciesla, and Lieutenant Colonel Zia Mehr for their friendship, guidance, mentoring, and support.

A number of people provided valuable and important comments on the manuscript. We are grateful for their help, though any mistakes are the authors,' not the reviewers'. Drs. Robert Bedoukian and Kevin Renskers were a tremendous help with chapter 8; Dr. Harold Harlan reviewed our work on bed bugs; Kevin Sweeney gave us a much more realistic perspective on the regulation of pesticides and repellents; Dr. Robin Todd helped with the evaluation of repellents; Rick Vetter reviewed our section on spiders, sharing a real passion for accuracy and science; Justin O. Schmidt provided an extremely helpful review of the entire book; and Dr. Graham White was a fount of knowledge on all aspects of medical entomology and bug control.

Finally, we would like to thank Peter Prescott and Tisse Takagi of Oxford University Press. Their responsiveness, advice, and encouragement were always welcome: they made this book possible.

PREVENTION OF BUG BITES, STINGS, AND DISEASE

WHAT'S BITING ME?

People naturally want to name whatever they see. We spend considerable effort during our lives learning the common names of everything from robins to rats, granite to marble, and pines to oaks. Unfortunately, there are more different things than anyone can know, and the words do not always keep up with reality. Bugs in common English can be anything from the flu bug to a tick to any insect. Bugs in entomological jargon are actually a single group of insects with straw-like sucking mouthparts originating at the front of the head, simple development, and front wings that are thicker at the base than at the tip. In this book, we will be less specific and just let bugs be bugs—any centipede, scorpion, spider, mite, tick, or insect that invades what we consider to be our space.

Most people are highly motivated to identify a bug that bites or stings. In some cases, the wrong name is applied so commonly that it becomes standard locally. That's okay among a small group of people who understand each other, but as soon as you want information from outside your local experience, you will need a name that is understood in the scientific community. Beyond curiosity, there are practical reasons to know the identity of what is biting. Once the problem is identified, it is possible to know how to prevent biting and the nature of the risks from the bites.

The rest of this chapter presents four different ways to identify biting bugs. It is usually easy to see what you need in order to get a practical identification; the challenge is that there are a lot of different things to see. It will be much simpler if you learn one technical term—arthropod. Arthropods include familiar things like crabs, scorpions, spiders, flies, and many other creatures. Think about animals with more than four legs, the legs with obvious joints formed from a thick, hard, or leathery exterior. That hard exterior (the cuticle) serves many purposes, much as our skin serves many purposes. However, the hard coating of most arthropods also serves as a skeleton, providing attachments for muscles, joints for limbs and segments, and structure for the organism as a whole. The size of arthropods varies from the microscopic to big marine forms. The majority of those you see on land are small, less than a foot (30 cm) long and seldom more than an inch (2.5 cm).

Arthropods can be divided into two major groups. One group is called mandibulate (with chewing mouthparts) and the other is chelicerate (with fang-like mouthparts). Among the mandibulate arthropods, the major classes are crustaceans and insects. Crustaceans are familiar to most people as gastronomic delights: crabs, shrimps, and lobsters. The group also includes barnacles, tiny water fleas, sowbugs, pillbugs, and a host of other organisms, usually aquatic. None of the crustaceans make their living by sucking human blood, though plenty are important parasites of other creatures.

Insects are seen by most people every day. The general characteristics of the group are a body subdivided into three main parts: head (front), thorax (middle), and abdomen (rear). There are two antennae (one pair) on the head and six legs (three pairs) on the thorax of an adult. Adults of many types of insects can fly, using one or two pairs of wings that are mounted angel-like on the top of the thorax. The common names of insects and other arthropods are important parts of the vocabulary of all languages. Although people might not be able to define a beetle, dragonfly, or caterpillar, they know one when they see one. Insects are a diverse group, occupying almost every ecological role imaginable. Many species suck blood to gain nutrition and, of course, many of those suck human blood. There are some unusual flies that have bloodsucking larvae (Congo floor maggot), invasive larvae (screwworm fly), or skin-infesting larvae (human bot fly), but they are not considered in this book.

We just mentioned some forms of insects that are special: caterpillar, maggot, and larva. These are representatives of one of the stages of the majority of insects that develop from egg to adult through a process called complete metamorphosis. The egg is an egg, varying in shape from a round, fluted object to a smooth, oval football. Eggs are laid (or, if you are an entomological purist, the egg is "deposited") in a variety of ways: singly or cemented in place or grouped together in small structures or injected into a plant or glued to another insect or injected inside another insect or placed in a wound or... The next stage is the larva (plural: larvae), an immature form of the insect that usually looks like a completely different animal. Figures 1.28 and 1.30 show the worm-like larvae of mosquitoes. The mosquito larvae, or "wigglers," live a completely different life from the adults, taking advantage of entirely different nutrients. Figures 13.1.1 through 13.1.5 show a series of caterpillars that energetically chew vegetation, compared to the moths and butterflies that only suck liquids and cannot even digest protein. The transition from larva to adult is extreme, requiring an intermediate stage called the pupa, which does not feed. Among the biting or stinging insects, the flies (including mosquitoes and midges), bees, wasps, ants, butterflies, and moths all develop through complete metamorphosis.

Other insects develop through a process called simple metamorphosis or, less correctly, incomplete metamorphosis. Plenty of changes occur during simple metamorphosis, but the basic body plan of the immature insect is similar to that of the adult—minus wings and sexual organs. You can see an example in the key to identification below; the adult kissing bug has wings and the immature kissing bug does not. Some insects with incomplete metamorphosis undergo much more radical changes, like dragonflies, which have aquatic predatory immatures and the familiar flying adults. Lice, bed bugs, and kissing bugs are the biting insects with incomplete metamorphosis.

Other mandibulate arthropods include millipedes, centipedes, and a couple of obscure little groups. None of these other groups suck blood, though some centipedes inflict painful, toxic bites.

The other major arthropod division is called chelicerate, because its members all have what are basically fang-like mouthparts. Most are arachnids, either with bodies divided into two parts (abdomen and head + thorax) or with only one body part. Arachnids lack antennae, and adults have eight legs. The class of arachnids includes many familiar organisms, including spiders, scorpions, ticks, and mites. Spiders might bite, but they do not suck human blood. Ticks, on the other hand, get all of their nutrition from blood. Of the many kinds of ticks, only a few species commonly bite people. Mites, as their name implies, are very small but otherwise rather similar to ticks. In contrast to ticks, mites make their livings in many different ways. Some are internal parasites (including some in people), others are predators, and many feed on plants. The human-biting mites are in a number of taxonomic groups. Often, these mites are so small that they pierce individual cells and suck out the cellular contents or they dissolve tissues with enzymes and take up the resulting liquid. Although this book only considers those mites that use humans as hosts for nutrition, certain mites that infest our homes and workplaces can cause severe allergic reactions of either the skin or respiratory tract. Some mites are intermediate hosts for a tapeworm that parasitizes people.

Other arthropods make up some of the odd bits of the invertebrate world, including sea spiders, horseshoe crabs, and tongueworms. The arachnids themselves have their share of minor orders, including daddy longlegs, wind scorpions or camel spiders, pseudoscorpions, and whip scorpions.

Concentrating on the arthropods that suck human blood, how can we identify the problem pest? The variety of mouthparts they use to get your blood is truly amazing, but they all get the job done with admirable efficiency (Figure 1.1). Matching the specimen to a picture can be helpful (Figures 1.2–1.44), but it often leads to misidentifications if you are not completely aware of the variety of organisms. For example, a person without experience might capture a harmless clover mite, see a picture of a rat mite on the Internet, and assume that he has the blood-sucking pest. Realistically, there are several approaches. First, you can rely on general knowledge and vocabulary (Table 1.1). This is a top-down approach in that you use the association of a word with an arthropod to communicate its identity, something usually learned during childhood or elementary education. Many people would be very accurate at applying common names, though the common

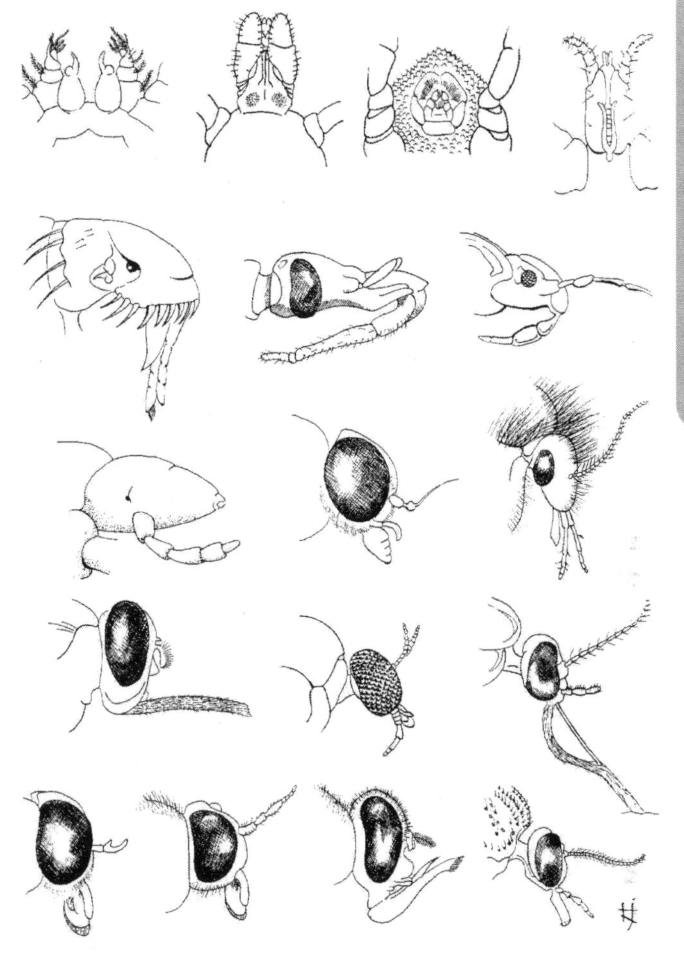

Figure 1.1. The wide variety of mouthparts used to suck your blood. *Left to right* and *top to bottom*: row 1 (ventral views): chigger, hard tick, soft tick, biting mite; row 2: flea, kissing bug, bed bug; row 3: louse, snipe fly, sand fly; row 4: tsetse fly, black fly, mosquito; row 5: horse fly, deer fly, stable fly, biting midge. See measurements in the identification key for an idea of scale.

names people know often refer to a group rather than to one particular kind of arthropod.

How about the bugs that sting or bite in self-defense or in defense of their colonies? The same general rules of identification apply and, for the most part, the common names are even more familiar (Table 1.2). Some of these arthropods have a rough resemblance to another group, causing some confusion in identification. In particular, biting centipedes and nonbiting millipedes have a lot of legs. The difference is that a

Table 1.1 Common Names Used by Most People for Blood-Sucking Arthropods

	Scientific Classification	
Common Names	**Order**	**Family**
Arachnids		
mite	Notostigmata, Holothydia, Mesostigmata, Prostigmata, Astigmata, Oribatida	many families
scabies	Astigmata	Sarcoptidae (species *Sarcoptes scabei*)
chigger	Prostigmata	Trombiculidae
tick	Ixodida	Ixodidae (hard ticks), Argasidae (soft ticks)
Insects		
body louse	Phthiraptera	Pediculidae (species *Pediculus humanus humanus*)
head louse	Phthiraptera	Pediculidae (species *Pediculus humanus capitis*)
crab louse	Phthiraptera	Pthiridae (species *Pthirus pubis*)
bed bug	Hemiptera	Cimicidae (species *Cimex lectularius* and *C. hemipterus*)
kissing bug, vinchuca, cone nose bug	Hemiptera	Reduviidae (subfamily Triatominae)
fly	Diptera	many
mosquito	Diptera	Culicidae
black fly	Diptera	Simuliidae
sand fly	Diptera	Psychodidae (subfamily Phlebotominae)
biting midge, sand flea, no-see-um, sand fly, flying teeth, black gnat	Diptera	Ceratopogonidae
snipe fly	Diptera	Rhagionidae
horse fly, deer fly, cleg	Diptera	Tabanidae
stable fly, dog fly	Diptera	Muscidae (species *Stomoxys calcitrans*)
tsetse fly	Diptera	Glossinidae
flea	Siphonaptera	many families

millipede has two legs per segment and each leg is generally much smaller, giving the animal an overall cylindrical appearance without a distinct head. Also, a millipede lacks the two terminal "tails" that trail behind the last segment of the body of a centipede and millipedes run much less rapidly than centipedes. Of course, centipedes bite and millipedes don't, but that would be a desperate way to distinguish them. Many people call yellow jackets "bees," even though yellow jackets are a kind of wasp. Bees have a much more furry appearance than do wasps. Finally, ants are commonly mistaken for termites

Table 1.2 Common Names Used by Most People for Venomous Arthropods

	Scientific Classification	
Common Names	**Order**	**Family**
Arachnids		
scorpion	Scorpiones	Buthidae and others
spider	Araneae	many
Chilopods		
centipede	Scolopendromorpha, Scutigermorpha, and others	many
Insects		
caterpillar, butterfly, moth	Lepidoptera	many
wasp	Hymenoptera	many
yellow jacket	Hymenoptera	Vespidae
hornet		
mud dauber	Hymenoptera	Sphecidae
spider wasp, pepsis	Hymenoptera	Pompilidae
ant	Hymenoptera	Formicidae
bee	Hymenoptera,	Anthophoridae, Apidae, Halictidae, and others

and vice versa. Winged ants have shorter wings, with the front wings longer than the hind wings; termites have wings that extend well past the end of the body and the wings are of approximately equal length. In general, wingless termites are very light in color and ants are usually darker. Also, the hind section (abdomen) of a termite body is broadly joined to the middle portion (thorax); ants have the classic "wasp waist" of their insect order, a narrow connection between the abdomen and thorax.

Another way to identify arthropods is to discuss what you have seen with a professional entomologist, often a mosquito abatement or agricultural extension person. This is a bottom-up approach, where the qualities of the arthropod are described in hopes of leading to identification. A digital picture or even a scanned image of the bug can help the entomologist come up with an accurate identification. Use the highest resolution (that is, the greatest number of megapixels) and be sure to get the focus right. Even if the image seems small, a good picture can be magnified on a computer by the expert, who might see the one part of the bug that makes for an accurate identification. In addition, below are some of the important things to tell a professional bug person.

Location where the bug was found:

- near light
- on wall
- on plant
- indoors
- outdoors
- in a place with stored products, straw, wicker, etc.
- on the arm, under the arm, on scalp, etc.
- in a nest underground, hanging from a tree, or in a wall
- under something like a rock or board

When the arthropod was found:

- time of day
- season
- weather conditions at the time it was found

What the arthropod was doing when it was seen:

- biting
- stinging
- attached to skin
- flying
- resting
- crawling
- hanging from a web

Its size:

- length in millimeters or inches
- comparison to household object or body part (for example, pin head, thumbnail)
- relative size (tiny, big, etc.)

Color:

- dark or light
- tan, yellowish, reddish, orange, black, brown

Markings on body or legs, for example, light-colored stripes

Shape and body characteristics:

- long and narrow
- round or oval
- flat
- long or short legs
- distinct head
- antennae

- pincers
- long tail with sting
- hairy
- hard and shiny
- with tiny scales

Number of legs:

- six, eight, or many

Shape of the mouthparts:

- long beak (sometimes tucked up under body or head) or not
- short and narrow
- paired fangs
- indistinct
- broad and triangular

Wings:

- none, two, or four

Color and transparency of wings:

- wings covered with scales or with some scales
- wings membranous and clear
- wings membranous but with "stained-glass" spots

Finally, there is the "foolproof" method often used by entomologists for identifying arthropods. This involves looking for a series of specific features of the specimen that unequivocally provide identification. Although simple in concept, this process often requires examination of the specimen following special preparation or with special equipment. The process can also require training or experience in recognizing the various features. That said, there is no reason that a nonprofessional cannot use the same process to identify the bug; even a preliminary identification would be helpful prior to discussing the problem with a professional.

The identification key (series of alternatives) below provides the information needed for accurate identification of biting and venomous arthropods (it's the KISS principle: Keep Identifying Suckers Simply). You should be able to use a simple hand lens to see all the features on the bugs, and the identification will lead to a name you probably already know. Bear in mind that most types of bugs do not bite or sting, so they are not included in this key.

WINGS

■ TWO WINGS

Wings held above abdomen while biting, hair-like scales on wings and body, small flies **SAND FLIES** (<1/5″ [<0.5 cm] long)

■ **Wings held flat against abdomen while biting**

■ **Many segments of antennae, small flies**

Long proboscis (beak or sucking tube), wings long and narrow, scales along veins of wings, scales sometimes forming light and dark spots **MOSQUITOES** (3/16″–7/8″, usually <1/2″ [0.5–2.2 cm, usually <1.3 cm] long)

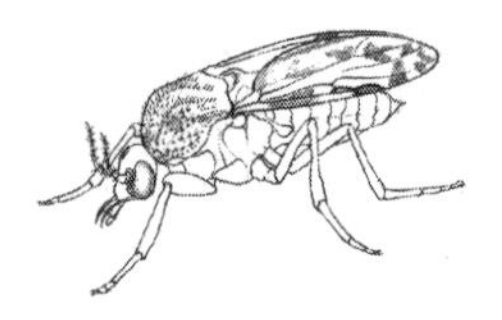

Small flies, rounded wings and short mouthparts, sometimes spots on wings but no scales **BITING MIDGES** (<1/8″ [<0.3 cm] long)

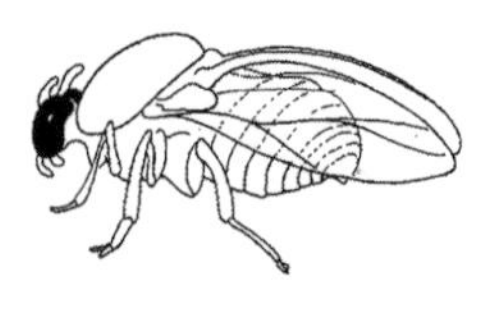

Small flies, usually very dark or gray with no or a few light markings, short antennae, hump-backed appearance, short mouthparts, wings broad at base, narrower toward ends **BLACK FLIES** (<1/6″ [<0.4 cm] long)

■ **Three segments to antennae, medium to large flies with stout bodies, wings clear or with stained-glass spots**

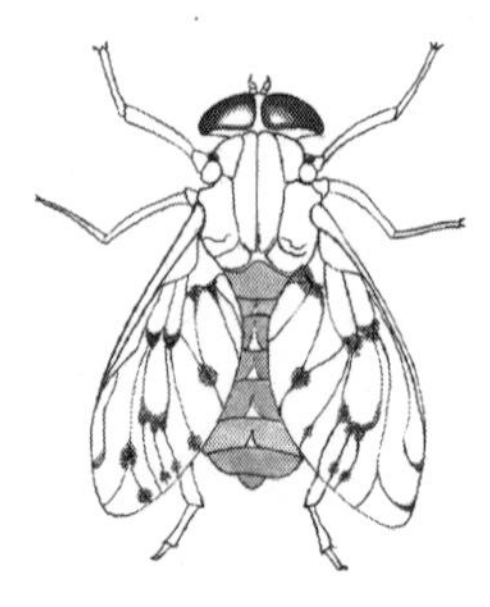

Third antennal segment long with rings around the tip, tip of wing with one vein on one side and another on the other side **HORSE FLIES** and **DEER FLIES** (1/4″–1 1/8″ [0.6–3 cm] long)

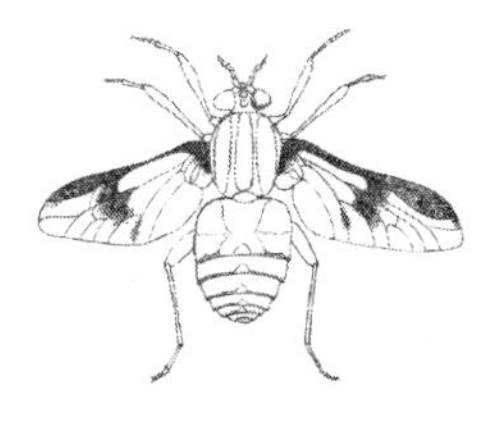

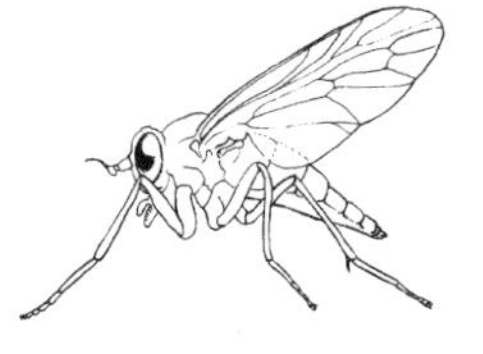

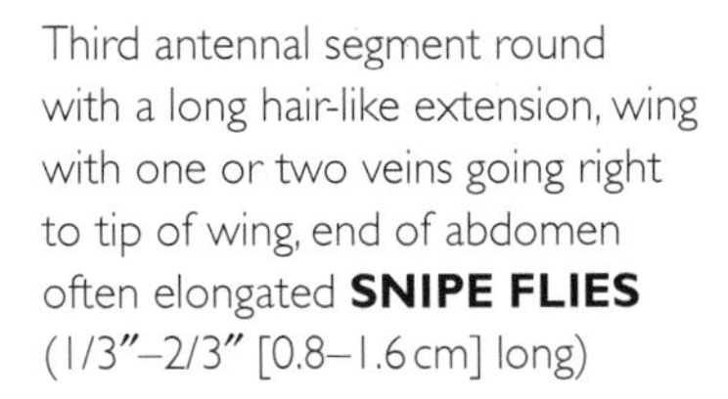

Third antennal segment round with a long hair-like extension, wing with one or two veins going right to tip of wing, end of abdomen often elongated **SNIPE FLIES** (1/3″–2/3″ [0.8–1.6 cm] long)

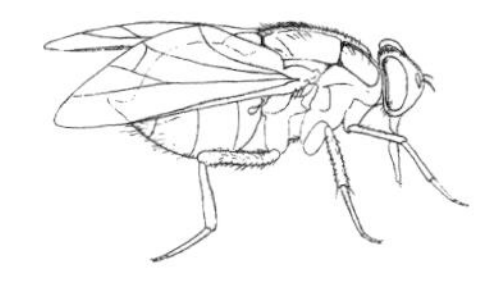

Third segment of antenna with a hair at the end, looks like a house fly with a bent mouthpart ending in a sharp point **STABLE FLIES** (1/3″ [0.8 cm] long)

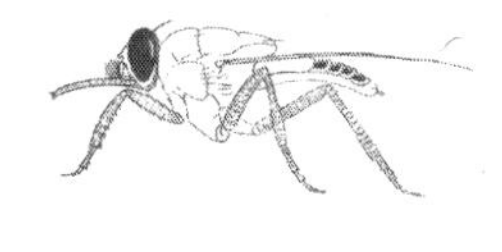

Third segment of antenna with a hair at the end, only in Africa and formerly in parts of the Arabian peninsula, long and pointed mouthparts, spotted abdomen, larger than a house fly **TSETSE FLIES** (1/4″–2/3″ [0.6–1.7 cm] long)

FOUR WINGS

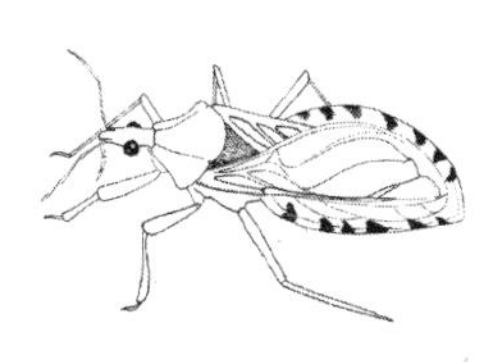

Wings held tightly against abdomen so that it is not clear how many wings are present, the top pair of wings thickened near the base and membranous at the tips, abdomen extending beyond wings at the sides and the abdomen curved upward at the edges, large insects, often marked with stripes and colors, head longer than wide with a groove between the eyes **KISSING BUGS** (1/4″–1 3/4″ [0.6–4.5 cm] long)

- Wings clear and thin, front wings longer than hind wings, thin connection between hind.part of body (abdomen) and middle part of body (thorax)

Body covered with hairs giving a furry appearance, first segment of hind leg flattened and expanded into pollen basket, connection between hind and middle body parts very short **BEES** (1/8″–1.5″ [0.3–4 cm] long)

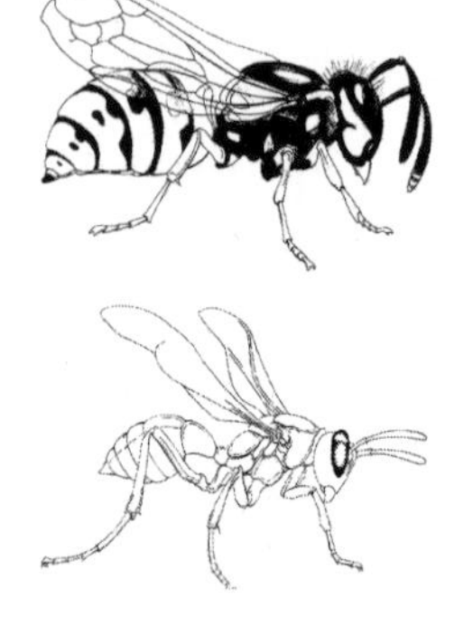

Body with fewer hairs giving it a hard, plastic-like appearance, often colorful in blacks, yellows, and reds, often a long connection between the hind and middle parts of the body **WASPS, YELLOW JACKETS, HORNETS** (1/2″–2″ [1.3–5 cm] long)

NO WINGS

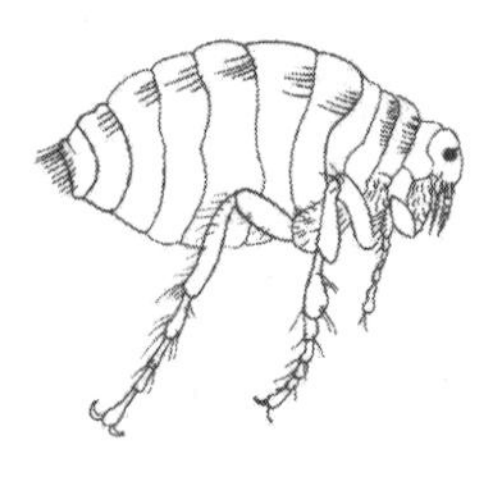

Flattened side to side, dark color, jumping insects **FLEAS** (<1/8″ [<0.3 cm] long)

- NOT FLATTENED OR FLATTENED BACK TO BELLY
- Six legs
  - Distinct head

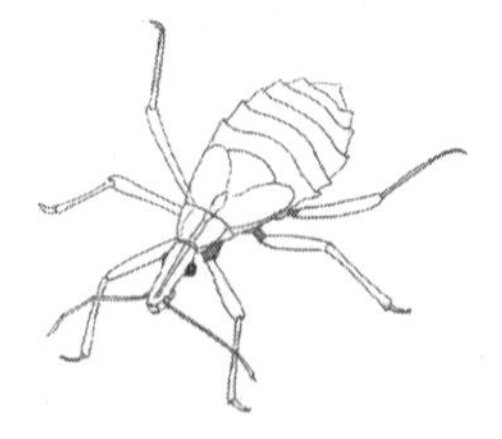

Various markings on body, thorax narrow, abdomen curved up at edges **KISSING BUG NYMPHS** (1/8″ to >3/4″ [0.3–2 cm] long)

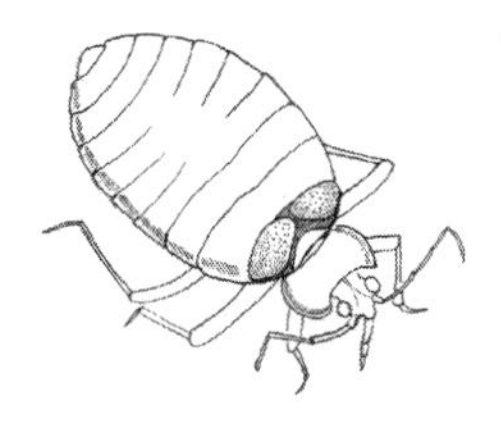

Cinnamon color (reddish brown), insect about as wide as long, eyes evident **BED BUGS** (¼″ [0.6 cm] long)

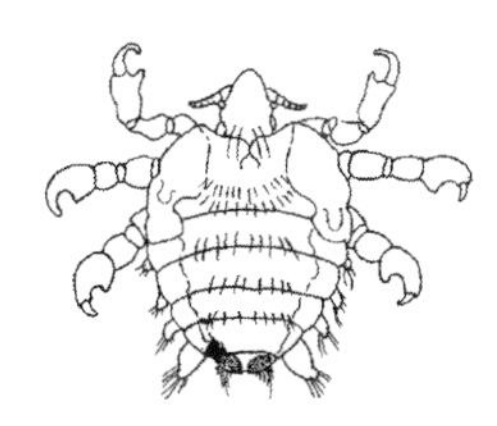

Light in color, eyes inconspicuous, usually in hair on head, armpits, or groin, sometimes in clothing **LICE** (1/10″–1/8″ [0.2–0.3 cm] long)

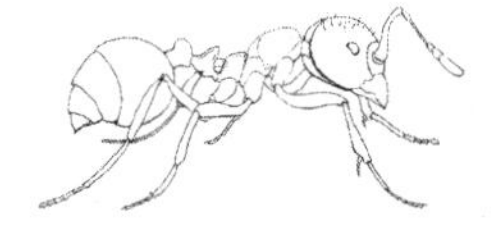

Varies in color, but usually dark, red, or yellowish, elbowed antennae, thin connection between hind part (abdomen) and middle part (thorax) of body **ANTS** (1/16″–1″ [0.2–2.5 cm] long) (velvet ants look like a furry ant, but are actually a kind of wasp)

■ Head not distinct

Appendages adjacent to mouthparts (palps) well developed, longer than mouthparts, often red, pink, orange, or yellowish, barely visible to the eye **CHIGGERS** (1/100″ [0.02 cm] long)

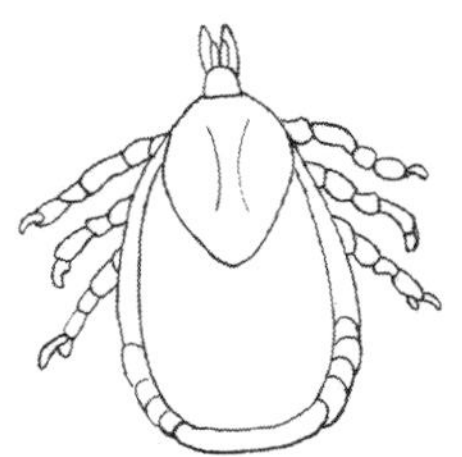

Appendages adjacent to mouthparts not well developed, shorter than mouthparts, often dark **TICK LARVAE** (usually about 1/16″ [0.2 cm] long)

■ Eight legs

Often flattened back to belly when not engorged, leathery texture, very short body hairs **TICKS**

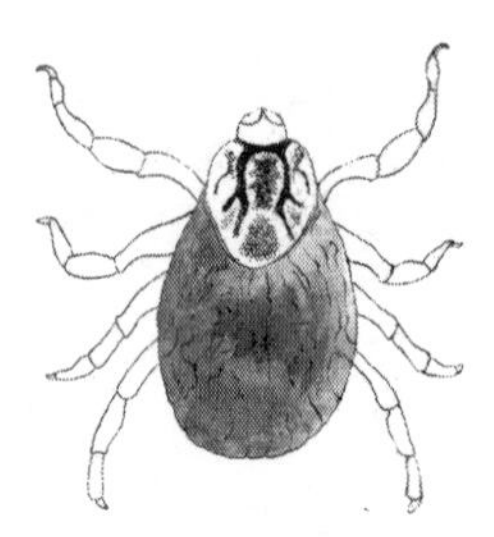

Distinctly flattened back to belly, attaches firmly to skin **HARD TICKS** (1/10″–1 1/4″ [0.2–2 cm] long)

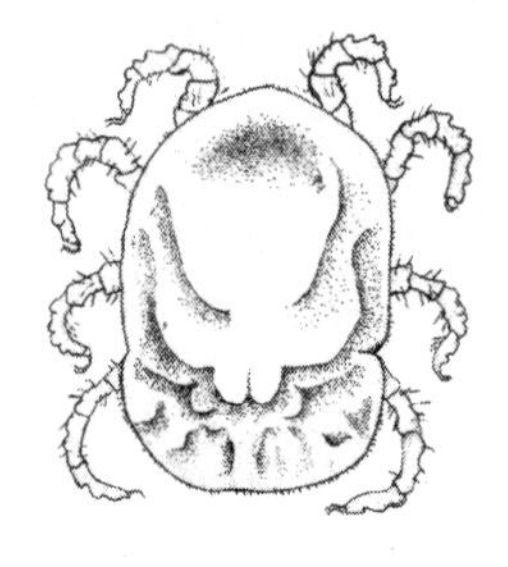

Less distinctly flattened, usually with small protuberances on body, completes feeding in a few minutes, often a painful bite **SOFT TICKS** (1/8″–3/4″ [0.3–2 cm] long)

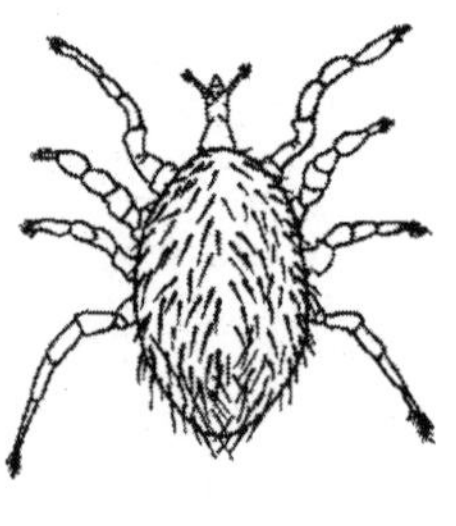

Not particularly flattened, often with long body hairs, body texture membranous with or without hardened plates **MITES** (<1/8″ [<0.3 cm] long)

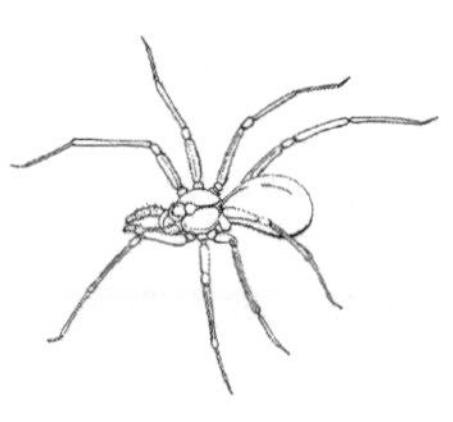

Two distinct body parts, hind body part (abdomen) without segments, produces silk, often constructs webs, mouthparts with fangs **SPIDERS** (1/8″–8″ [0.3–20 cm], including legs)

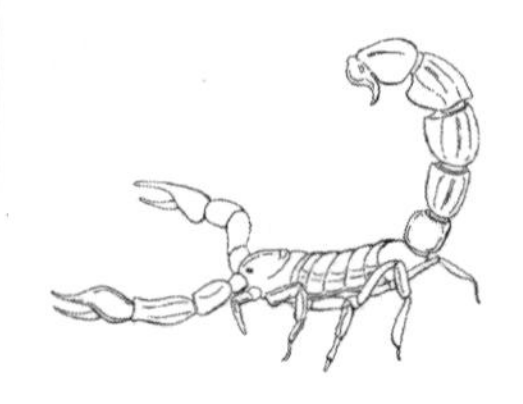

With crab-like pincers, segmented body, and long tail ending in stinger **SCORPIONS** (1/2″–8″ [1–20 cm] long)

Many legs, distinct head

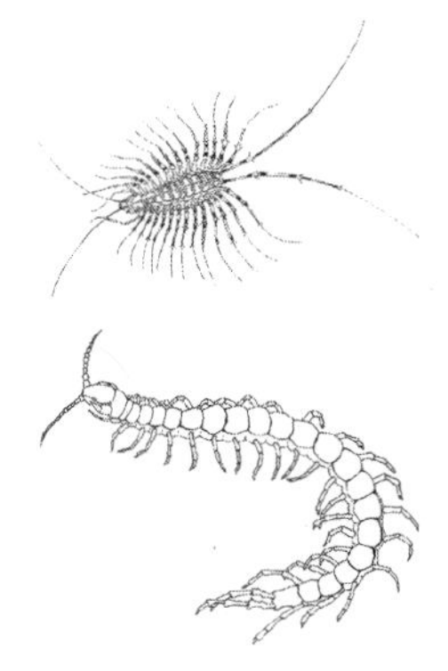

Many legs with one pair per segment, distinct antennae, poison claws behind head
CENTIPEDES (1/2″–12″ [1–30 cm] long)

Identification of a biting arthropod from the feel and appearance of a bite is an inexact science at best. One of the problems is that each person feels something a little different and each one's skin reacts in a different way. Also, the response of an individual to bites changes as the immune system adapts to repeated exposure. Sometimes, scratching the bites can lead to secondary infection of the skin, resulting in even more variation of how the bite looks. Some individuals develop an allergic response to bites or stings from any of the arthropods, varying from swelling to fatal anaphylactic shock. The following remarks may be helpful in identifying some bites.

Mites: There are about 10 species of mites that bite without burrowing into the skin. These bites vary, but they have several things in common. First, they tend to itch badly. Unlike an itchy rash, the bites are distinctly separated, often with a raised area and a halo of reddened skin. Biting mites never leave a visible puncture where they bite. Scabies mites are a special case. Scabies is an infestation best treated under a physician's care. This condition is dreaded by anyone who has had it or by family members who have had to listen to them. The mites literally burrow through the upper layers of the skin at three separate points in their life cycles. Eventually, a person develops an allergy to the eggs, feces, and mite parts left behind in the skin. The reaction causes unbelievably intense itching at night and a weeping rash similar to poison ivy even on uninfested parts of the body (Box 1.1).

Box 1.1 Scabies: The Mighty (Miserable) Mite

You notice that your wife (who had been using borrowed gloves at her garden club) has developed an annoying habit. She started by scratching her hands and arms for a half hour at a time. Lately, she almost never stops and even wakes you up at night as she rubs the slightly inflamed skin over and over. It has been going on for weeks, doesn't get any better, and is really wearing her down. Finally, she goes to her primary care provider, who fortunately has had wide experience with the elderly, the indigent, and the recently arrived immigrant. The physician recognizes the problem right away from the more intense irritation and slight scabbing on the webs of the fingers, the complaint that the itching is much worse at night, and the appearance of faint gray lines under the skin of the hands. He knows that the patches of weeping irritation on other parts of the body are a side show, not the main problem. For this patient, it won't be necessary to get positive proof of infestation by attempting to dig out a mite from one of the burrows. He also knows that he had better get this person away from his other patients and staff, and he gently directs her to the examining room he uses for potentially infectious people. Washing his own hands immediately and thoroughly, he gowns up carefully before greeting the woman again to inform her that she has scabies.

Scabies is caused by a mite that is almost too small to see and that is almost never noticed by the person who suffers with an infestation. Other animals have their own species of mites that cause conditions often described as mange, but the human scabies mite is restricted to us. The mite burrows just under the skin, leaving a trail of eggs and feces. Larvae emerging from the eggs go to the surface before burrowing in again. Females often emerge from the skin and burrow in again. The mites can get quite numerous, but even a few can lead to the severe symptoms described in the case above. That's because the scabies mites have a damaging trick up their sleeves: not only do they damage the skin with their burrowing, but they also leave behind proteins that cause an allergic response similar to poison ivy. After about six weeks of an infestation, many people start getting patches of intensely irritated skin on their legs, face, or other parts of their bodies that are not necessarily infested with the mites.

Typically, the mites start in the hands or groin, but they can live on skin anywhere. Intimate contact between a person and their scabies mites sets up an immune response that limits the infestation in healthy people, but an untreated case of scabies can last for many years. A few unfortunate people fail to mount an immune response, and the mites multiply to a level of many hundreds of thousands, eventually killing the victim. A person with scabies is highly contagious through direct contact and also by sharing mites that get rubbed off onto bedding, clothing, or gloves. At warm temperatures (greater than about 75°F or 25°C), these mites don't live more than a few days off the host; however, cool temperatures can extend the mites' survival for a couple of weeks.

Scabies is a medical condition that can be cured with a prescription ointment containing 5% permethrin. The usual recommendation is to apply the ointment head to toe after showering in the evening, and then shower it off again in the morning. This treatment needs to be repeated after about 10 days to kill any mites that have hatched from eggs. Simply following these directions does not usually eradicate the infestation, however. It is pretty disappointing to get some relief for a couple of weeks, only to see the symptoms reappear. One problem is that other people in the household may be infested, but not yet have the allergic response that causes most of the irritation. The other problem is that even a single female mite in bedding or clothing can start the infestation all over again. You can be more certain of success if you treat everyone in the household at the same time and separate them from any unwashed (washing in hot water and heat drying kill all mites) clothing, unwashed bedding, or mattresses that they used before the treatment. This is a tall household order, sometimes requiring that people move out of their bedrooms and use sleeping bags for a week.

Although perfectly healthy people can get scabies, it is much more common when there is another medical condition. Tragically, scabies is common among the homeless, the elderly in care facilities, and, malnourished children. The problem is almost unknown by those fortunate people who live healthy lives under good economic conditions. Be grateful.

Chiggers: Chiggers in the Americas cause some of the most irritating bites in the region. They secrete a proteinaceous tube within the skin that causes an allergic reaction. Long after the chigger is gone, the skin continues an immune reaction to the tube, causing intense itching for weeks. The bites appear as irregular red spots on the skin and are usually clustered in areas where the clothing is tight against the skin (belt line, top of the socks, under the bra, etc.).

Hard ticks: Ticks feed by penetrating the skin and then secreting substances that cement the mouthparts into the surrounding tissue. Some species secrete saliva that can be described as toxic. In some cases, a tick located above the upper part of the spine can cause paralysis of the entire body. The feeding process goes through two major stages, first preparing the site and then imbibing a large amount of blood. It is during the imbibing stage that most pathogens are transmitted. Therefore, early detection and removal of ticks can prevent most transmission of the bacteria and viruses carried by ticks. The site of a tick bite often becomes irritated, but the results are variable.

Soft ticks: One species of soft tick, the pajaroello tick of California and Mexico, has a bite that is described as the most painful in the arthropod world and similar to that of a rattlesnake. Bites are fairly frequent among people who work or camp outdoors in drier wilderness areas. Fortunately, the bark of the tick's reputation is not as bad as its bite. Very few people experience more than mild irritation and a small reddened swelling that lasts a few weeks.

Lice: Louse bites are more irritating than you would expect, considering that these insects live on humans for months or years. Hair usually hides the bites, which are generally flat and reddened with a tiny central dot. Exposure over a long period can lead to desensitization, though extreme exposure causes skin thickening and discoloration. Crab (pubic) lice are said to leave a strange blue spot on many people. All lice cause a constant sensation of itching.

Bed bugs: An individual's response to bed bug bites can vary from almost no response to a generalized skin irritation that extends all over the body. Some people get large welts or even blisters. Most commonly, the bites are similar to those from mosquitoes. Many bites occur on the head, upper body, and arms. Continuous exposure to bed bug bites can result in a generalized rash that is difficult to distinguish from other sources of skin irritation. Often, a person will notice some irritation and difficulty sleeping, but not recognize the fact that dozens of bugs are feeding every night. The bites often occur in a linear group caused by the bugs' feeding behavior. These rows of bites are the result of the bugs' frequent habit of standing on a different surface (like the edge of the sheet) next to the part of the body on which they are feeding. The straw-like mouthparts line up along the edge, like a row of people sipping piña coladas at a bar.

Kissing bugs: Some of the species of these bugs inflict painful bites, but most others are nearly painless. Some people develop significant irritation following bites and this can enhance the chance of acquiring an infection with the parasite that causes Chagas disease. The dangerous species of kissing bugs deposit parasite-laden feces while they are feeding. As a result, people who scratch the bite may push the parasites right into the wound or into their eyes.

Mosquitoes: The responses to mosquito bites have been said to go through five immunological stages. Each species has different components in its saliva, though there appears to be some cross-reaction between related species. When you are exposed to a new group of mosquitoes, you may not have much reaction at all. Subsequent bites begin to cause a primary reaction characterized by a pale wheal around the bite. The wheal continues following more exposure, with the addition of an itchy red swelling that may last for several days. As the immune system adapts, the wheal stage may disappear completely, but the itchy red swelling remains. Finally, after much exposure, the reaction becomes distinctly milder. Some people claim to be particularly sensitive to mosquito bites. Probably some are actually very sensitive, but others may be at the stage where they are sensitized, but not accustomed. Mosquito bites often occur in a line, as the mosquito probes repeatedly before settling in for a blood meal.

Sand flies: Possibly because most people have little experience with these flies, they can cause a very irritating bite that continues to itch for over a week. The bites often seem to scab over, creating the appearance of chicken pox when they are numerous.

Black flies: Most black flies like to bite in hairy areas, like the edge of the hair of the scalp. They can sometimes crawl around for a long time before biting. The bites often have a red center. Some individuals have a particularly bad reaction that eventually results in permanent scars.

Biting midges: At least some biting midges cause a distinct lesion that is pancake shaped with a red dot in the center. One group of biting midges is particularly irritating, biting around the ears and on the head, sometimes causing such a bad reaction that people feel ill and may even develop a fever.

Horse flies, deer flies, snipe flies: These flies have mandibles shaped like scissor blades. They slash the skin and then lap up the blood that pools on the surface. The long-term effect is minimal, but the immediate effect of the bite is often painful.

Stable flies: Stable flies are very persistent biters, causing a sharp pain as they pierce the skin. The lesion is itchy, but otherwise unremarkable.

Fleas: The common fleas cause lesions that are typically round, domed in shape, and evenly reddened in color. The bites flatten over time, leaving red spots. Often, the bites are concentrated on the lower legs. Some people think flea bites are particularly itchy.

Venomous arthropods: The reaction can vary from a little red bump to a fatal systemic reaction. See chapter 13 for details.

Now that you are aware of the variety of biting bugs, the next question after "what" is "where."

LOCATION, LOCATION, LOCATION

Identifying the arthropod that is biting you is very important in the process of solving the problem. For example, if it's a mosquito, you might think about cleaning up trash, unblocking roof gutters, or taking other action that eliminates standing water where mosquito eggs are laid, hatch, and develop. If it's a tick, you might treat the area with insecticide or keep the grass cut short. If it's a biting midge or black gnat...well, the solution isn't simple.

But what about avoiding the biting arthropod completely? A person can simply not be in the same place as the arthropod. Stay out of Africa, and you will not be bitten by a tsetse fly. Stay away from swampy trails, and you are staying away from deer flies. Sleep on the top bunk, and some types of mosquitoes are more likely to attack your partner on the lower one.

MITES

The mites that bite are closely associated with their principal hosts, whether rats, mice, pigeons, rabbits, or other kinds of animals. It should not seem too surprising that mites are more likely to occur where people carry these animals with them or when we create conditions that favor the mites' hosts. Perhaps the best example is the rat mite. It may have been restricted to Southeast Asia before people carried its rodent hosts all over the world. Now it would be hard to find a place with people that does not have these mites. Some regions have more of a problem than others, like San Francisco Bay

Area of California. Not so much in spite of, but more because of a generally prosperous human community, rats flourish in this area of abundant resources. Because of the mild climate, homes often have features largely open to the outdoors, creating easy access for the rats. In addition to the usual sources of food for rats in any city, the Bay Area also has an abundance of rat food in the form of backyard trees that produce fruit during much of the year. So the conditions in this area that cause a greater problem with rat mites are a chain of cause and effect: location and history have made a great place for people, people have made a great place for rats, rats prosper near and in people's homes, and rat mites escape from rats to make the place a little less great for people.

Even within a region where mites are common, there will be places and situations where they are absent and other places where bites are almost inevitable. Some homes are built in such a way that rats have multiple points of entry. Stout hardware cloth or other measures can stop them, but sometimes it is difficult to block all the openings. Consider a tile roof laid above a plywood underlayer. The rats can enter at the end of any of the hundreds of tile ends and then make their way through gaps in the plywood (some of which they might make themselves) to the attic and the rest of the house. Poor repair can open up a home to rat tenants. Old commercial properties may have gaps and holes, especially in wooden walls damaged by moisture or termites.

Is it possible to avoid rat mites within an infested home? Rat mites are often hidden behind walls, in basements, or in an attic space so that their source is not obvious. People who suffer with rat mites often make two comments that might be a lesson to the rest of us. First, only one or two people in the household are getting the bites (solution: don't be that person). And second, bites occur when that person is in one particular place in the house (solution: don't go to that place).

Fur mites are another interesting example of biting mites. These mites are smaller than rat mites and lighter in color, and their individual bites are often not very painful. As a result, they may go unnoticed on the rabbits, dogs, and cats that are their normal hosts. They become abundant in the scurfy skin flecks on the pet, which is the origin of their charming moniker, "walking dandruff." Fur mites occur all over the world, but they are most common where there are many pets. The worst problems from these mites occur when people have habitual close contact with their pets. Although the mites complete their life cycles on the animal (complete with eggs attached to individual hairs by cocoon-like silk), they sometimes move short distances away from the host, causing infestation of bedding or cages. Avoiding those locations and direct contact with the pets might be a short-term solution to prevent bites on humans. In one poignant example, a little girl was plagued by fur mite bites for months until one day she decided to take a nap in the box where her pet dog always slept. Shortly after, her parents were alarmed to see hundreds of bites on her limbs, which finally led them to find the cause and to have the dog treated.

CHIGGERS

Chiggers do not seem to distribute themselves around the world in the same way as some biting mites. Each of the dozens of species of chiggers remains in its own geographic range, though some of those ranges are huge. Chiggers may be tied to certain localities because they have a life cycle that involves much more than their mammal, bird, or reptile host. The larva that comes right out of the egg is the only stage that bites. Those motivated toward linguistic precision call the biting stage the "chigger" and the other stages, collectively, chigger mites. The next two active stages (nymph and adult) live in the topmost layer of the soil, where they are predators of arthropod eggs. Each species has its own favorite hosts, climate, and soil type, which seems to stabilize where in the world they occur.

So where are the best places for chigger bites? The itchy rash concentrated around the belt line and lasting weeks is pretty much an American phenomenon, especially in the midwestern United States but also in the Southeast and down into tropical America. Itchy chigger bites are not unknown outside these areas of the Americas, but the problem is much more limited. Elsewhere in the world, one hears about irritating chiggers in Europe, Africa, Australia, and, to a lesser extent, Southeast Asia.

There are few places in the world, except deserts, where you can completely escape these arthropods. A different group of chiggers causes much less irritation, but transmits the bacteria that cause a serious disease, scrub typhus. These chiggers prefer to feed on rodents, and a person usually receives only a few bites that may hardly be noticeable. They occur most commonly in southern Asia (as far west as Afghanistan), eastern Asia (as far north as the Primorye region of Siberia), Southeast Asia, Australia, and the Pacific islands. One species has been shown to ride on birds migrating between Japan and northern Thailand.

Once in chigger country, we find these mites to be notoriously spotty in their distribution. Much has been written about the type of habitats favored by chiggers, especially where scrub typhus is a problem. Whether it is lalang grass, oil palm groves, blackberry thickets, or river bottoms, the common theme is abundance of the primary hosts. Bring in big numbers of the right rats, harvest mice, or lizards, and you are far more likely to have chigger problems. When conditions are right for chiggers, they tend to concentrate in "mite islands." Again, this is probably as much driven by the movement of the host as by the movement of the mites. The larval chiggers will drop off more often where the host spends most of its time. Homeowners can get some relief by thinning vegetation or removing food sources for hosts, but people usually get chigger bites away from home when they are spending a lot of time outdoors. Nothing seems to help in the midwestern United States in the summer, and you might get a few bites even while sitting on a well-groomed lawn.

Chiggers are small and cannot fly or jump. Although they move remarkably fast (often escaping from laboratory colonies), the larval chiggers generally rely on the host to come to them. Biting is the only mission in the larval stage, so they are innately good at positioning themselves where they can climb aboard. This location may be at the tips of vegetation, where a lumbering human or rat might brush the mite onto itself, or it might be the soil where the host is likely to rest. This aspect of chigger behavior is the basis for the standard advice to avoid sitting or lying directly on the ground where these mites are abundant. At least in the United States, somehow the best places to rest (like embankments and bare sandy hummocks) seem to be the worst places for chiggers. Coincidence? We think not.

TICKS

Ticks have done pretty well for themselves. They have been residents of this planet since at least the Mesozoicera, right alongside blood-sucking insects. There are about 650 species of hard ticks and 170 species of soft ticks worldwide, covering a number of lifestyles and attacking many kinds of hosts. Geographic ranges of important species are usually restricted to a particular part of the world, though there are a few examples of ticks that have traveled everywhere thanks to human trade in animals. As a result, the United States and other countries spend considerable effort trying to stop certain tick pests of cattle from crossing their borders.

So where in the world can you go to avoid ticks? The answer is nowhere: there are ticks even in Antarctica. Of course, some areas are worse than others. For example, the Republic of Korea (South Korea) with its decimated mammalian fauna seems to have much less of a problem with ticks on people than many other countries with similar climates and topography. On the other hand, it is hard to avoid tick bites in parts of the United States where deer are abundant. Presumably, these distributions are related to the abundance of the right hosts for the ticks that happen to live in that area. Host requirements can be a bit complicated because many ticks feed on small animals when they are younger and larger animals when they are older. People get caught in the middle of this spectrum, often being attacked by younger and older ticks of a given species. A worldwide tour of important ticks would have to make quite a few stops. Canada and, especially, the United States have a wide variety of ticks that commonly attack humans, though no single species is distributed across all of North America. Soft ticks are more of a problem in the West than in the East. The common human-biter in Latin America seems to be more widely distributed in that region, though this may be the result of poor documentation. Europe has its own problems with an important soft tick in Spain and Portugal and a troublesome hard tick distributed from Britain to Russia. Things get really interesting in Africa and Asia, where there are whole groups of important hard and soft ticks affecting people in the many millions from north to south and from east to west. Australia has its worst problem along a narrow band of tick trouble along its east coast, which happens to be where most people live.

There are certain types of places where you are more likely to get bitten by ticks. "Ticky" places are often associated with the movement of rodents or larger animals. Ticks seeking a blood meal gather along such paths on the ground or at the tips of vegetation, their front legs outstretched to expose their sensory organs, ready to grab onto a host. The path might be a trail or road designed for humans with the ticks particularly concentrated in the vegetation along the edge. Even where there has been no human activity, hikers are likely to follow the lines of least resistance, which is exactly where many other animals have passed. Another good way to encounter ticks is to visit a water hole. Animals follow openings to the water's edge where the slope is gentle and there is good access for drinking. People almost inevitably want to gaze upon the water's edge, and they are likely to follow the same paths as the animals. Returning to the car, they might find hundreds of ticks headed for every crevice on their bodies. Sometimes infestations are so heavy that it doesn't seem to matter whether you follow paths or not. Such areas might be the result of a particularly productive year for the ticks or the result of the concentration of many wild animals around a scarce water hole or other limited resource. For the individual who finds a tick on clothing or body after a walk in the woods, there is often the impression that the tick fell out of the trees. In fact, this almost never happens, and you are usually correct in looking downward for the source.

Within a ticky area, ticks are most likely to be present in places where their relatively delicate bodies are protected from drying. As a result, ticks are less likely to stick around in areas that get hot and dry during the afternoon, such as locations with no vegetation or only short vegetation. If you can avoid scrubby areas where stems are constantly brushing against your legs, you may go a long way toward limiting your exposure to ticks. Also, sitting or lying down in tick country can make matters much worse. Soft ticks like places where animals sleep, old cabins, and caves. Be careful not to put your body where the residents are waiting for their dinner.

BED BUGS

A bed bug (either male or female) imbibes about 8 milligrams of blood during the 3–12 minutes that it feeds, meaning that it would take about 625,000 bugs to drain you completely. Each bug bites about once per week and lives for three months to a year, depending on temperature. Unfortunately, bed bugs can live a long time without feeding—even up to a year. A female bug lays 300–500 eggs during her lifetime. At that rate, one pregnant bug would produce enough offspring in about two years to exsanguinate a single sleeper. Fortunately, not every baby bed bug lives long enough to reach its full potential. A host of other household inhabitants feed on bed bugs, including mites, pseudoscorpions, spiders, ants, and even other bugs in the same taxonomic order (one of which is poetically called the "masked hunter," or *Reduvius personatus*). Bed bugs also have their own diseases, and a certain number of females succumb to the attentions of overzealous males that inseminate their partners directly through the female's abdominal wall. Ouch!

Bed bugs are the perfect biting insect. They do not transmit any diseases, but they live right in our bedrooms and in hotels. An individual bed bug makes itself at home by biting the sleeping human occupant every week. It is absolutely amazing that a room can be loaded with hundreds of the creatures, but the resident may not know why he cannot sleep well, why he has a rash on the back of his neck, and why his sheets seem to get spots on them. There is one common species (*Cimex lectularius*) in temperate areas that probably originated in the Middle East and now lives everywhere there are people. The other species (*Cimex hemipterus*) is more tropical, but its origins are unknown. Taken together, there is nowhere on earth where bed bugs cannot occur along with humans, because they require exactly the same conditions as people.

In spite of the wide distribution of bed bugs, some areas have more trouble with them than others. They seem to be a particularly common problem in the Middle East, for example. Still, if you want to estimate the risk of being bitten by bed bugs, you should look more at local conditions than geographic location. Areas where people frequently come and go are obviously susceptible. The bugs are good at hitching a ride to a new location on luggage or clothing. Many hotels and motels are infested, and you might be surprised by the frequency with which even the most luxurious places are forced to treat their rooms for bed bugs. In parts of the world where people might not have the means to purchase insecticide, bed bugs can get completely out of control. Under such conditions, the insult of anemia, insomnia, and severe allergic reactions in the inhabitants may be added to the injury of large reservoirs of bed bugs, which make transmission to other sites all the more likely.

Once you are located in the same room as bed bugs, you are likely to get bites near the head, on the upper arms, and on the chest. There are some signs of infestation that might be noticed even while the bugs are hiding in mattresses, furniture, and behind wallpaper. First, the temperate species leaves a sweet, almost pleasant odor most noticeable in long-established or very large infestations. Some individuals cannot detect this odor, which is perhaps not so surprising considering how our sensitivity to odor depends so much on our surroundings and experience. Pest control firms increasingly take advantage of the presence of bed bug odors by using specially trained dogs (see the National Entomology Scent Detection Canine Association, www.nesdca.com) that can sometimes alert them to the presence of even a single bug. Second, the bugs have the unlovely habit of leaving spots of feces (formed from your own blood) on the walls, furniture, and linens. Finally, it is sometimes possible to see the distinctive little creatures, or their yellowish, papery cast skins, by looking under drawers, in the seams of the mattress, between the mattress and the bed frame, or in sticky traps. Sticky traps can get bed bugs, though they sometimes fail to catch the insects even in heavy infestations. The expression "Don't let the bed bugs bite" evokes visions of a time when these blood suckers were a universal feature of the bedroom. Ah, the good old days.

KISSING BUGS

The "kiss" of kissing bugs is the kiss of the vampire: some species that attack humans tend to bite around the mouth while a person is sleeping. It is just as well that bites usually occur painlessly during sleep because people might be disturbed to see these 1–2″ (2–5 cm) critters when their abdomens are full of blood. Most kinds of kissing bugs are content to feed on animals far away from homes and therefore never encounter humans. The problem species like to come indoors, either living in crevices and behind furniture like bed bugs or flying into the house at night. One of the interesting aspects of these bugs is that they are the occasion for mentioning a style of feeding in which younger bugs take blood from a larger bug while it is feeding. Some enterprising entomologist called this *cleptohaematophagy*—a 50-cent word if there ever was one.

With one species exception, kissing bugs only live in the Western Hemisphere. The only species found worldwide followed domestic rats as they were distributed inadvertently by humans. Kissing bugs occur as far north as the central United States and south into Argentina, the kinds of species changing gradually from north to south. Some of the species have habits that put them close to humans much more often. In the United States, the species seem particularly tied to certain rodents and other nesting animals so that people usually only get bites when they occupy an old cabin or other location with wildlife activity. Farther south, particularly in Central and South America, a few species infest homes that offer a lot of crevices, cracks, and holes. The thatched, mud-walled homes of some rural people make excellent homes for kissing bugs as well. A person visiting the infested regions can avoid much of the risk by staying in quarters with smoothly constructed walls and ceilings. In fact, a major part of a Brazilian campaign against these bugs was plastering over rough, mud walls. Another risk factor is the presence of animals in the house. This is not uncommon in rural South America, where baby chicks are protected from predators by rearing them in a bedroom or guinea pigs run about loose as cute, edible pets. If you find yourself in a room that might be infested, use a bed net and push the bed away from walls and animals.

LICE

Three species of lice infest humans. They live only on human beings. Other animals and birds have their own types of lice, which cannot survive on us. Among people of all races, there are head lice (*Pediculus humanus capitis*) infestations that sweep through schools as children play games with a lot of contact or when they exchange clothing. Then there are crab lice (*Pthirus pubis*) that like thick hairs on eyebrows, in armpits, and in the groin regions of the human body. Finally, there are the real bad boys of the louse world, the body lice (*Pediculus humanus humanus*), which have been responsible for some of the worst plagues in human history (see chapter 3). Body lice live and reproduce in clothing, coming out to bite when they feel like it. All three of these lousy types are exquisitely adapted to people; we are the only host species they infest. Lucky us.

Being human parasites, lice have pretty much gone wherever people have gone. Although there is no record we have found, lice have probably followed people even to Antarctica. Some regions have more trouble with lice at one time or another. For example, the United States started suffering major outbreaks of head lice in schools in the 1970s. Parts of Africa have had terrible problems with body lice since the 1970s, with a major outbreak of louse-borne typhus in Burundi in 1995–1996. Crab lice can be the almost inevitable result of close physical contact between promiscuous partners. If there is any generalization to be made, the adage to "follow the money" applies. In this case, follow the money away from lice because they thrive on conditions where nutrition and sanitation are less than perfect. Body lice, in particular, thrive on human suffering, as happens during famine or war. Even the healthiest person who gets exposed to any of these lice is likely to develop an infestation. The threat is not really regional, but it is extremely local. Share the wrong bed, exchange clothing with the wrong person, or borrow the wrong earphones, and you are likely to get an infestation of lice that you will then share with others. That's how they get around.

MOSQUITOES

Everyone knows about mosquitoes; the word is almost as familiar as "cat" or "dog." People usually know a lot about where mosquitoes are abundant and when they bite. Most people also know that mosquitoes are associated with water, though relatively few would recognize the worm-like, aquatic larva. The public is usually surprised by the big variety of mosquitoes (there are close to 4,000 species, and new ones are described every year). Curious people ask two questions with surprising regularity: "How long does a mosquito live?" and "Why are there mosquitoes?" (Better to ask, "How long until mosquitoes die?" The answer is 10%–40% per day for many species that have been studied. As to why there are mosquitoes, a personal favorite answer is "because there can be.")

The worldwide distribution of mosquitoes follows patterns that are typical of other groups of organisms: more species in the tropics, some evolutionary lineages (presumably older) occurring around the world, and other lineages (presumably younger) restricted regionally. That said, not all mosquitoes were created equal in terms of their capacity to spread disease or simply to bother people with their bites. In fact, our association with mosquitoes is close enough that many of the worst species have been favored by human activity. For example, the yellow fever mosquito puts its eggs in containers just above the water line, waiting for either filling or condensation to hatch them. The adult mosquitoes do not fly far, and they love to bite humans. As a result of this suite of habits, they do very well anywhere in the world where people store water or where water is carelessly allowed to accumulate in discarded material that holds water, commonly bringing with them the threat of dengue virus (Figure 2.1). It is a good thing that this species does not survive a cold winter, though it often gets introduced to northern locations for the summer. Within historical time, people have carried this mosquito to many new locations, where it finds everything it needs in the domestic environment. It is probably not a coincidence that several important pathogens of people are associated mainly with this species. The distribution of the yellow fever mosquito

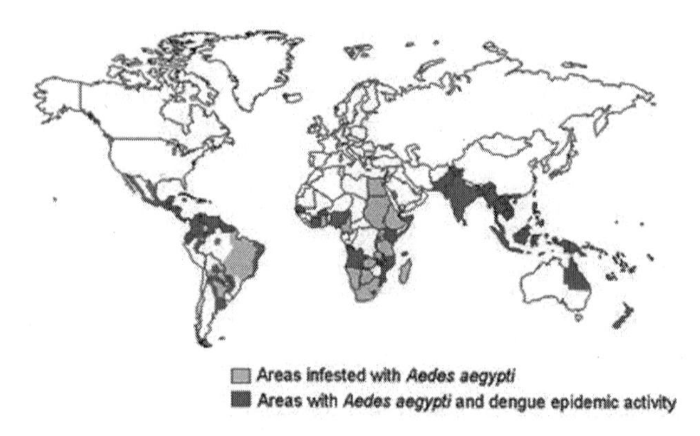

Figure 2.1. Distribution of the yellow fever mosquito (*Aedes aegypti*) and one of the viruses it transmits, dengue; based on 2005 data. Courtesy of the Division of Vector Borne Infectious Diseases, Centers for Disease Control and Prevention.

is an extreme example of the inadvertent domestication of a mosquito, but other species are also favored by human activity. It is only logical that, of the many species out there in nature, the ones that get blood from people and that lay their eggs in human-created sites would tend to be numerous near habitations. As time goes on, those species get better and better at exploiting those nonnatural habitats close to the huge source of blood provided by a population of 6 billion humans.

If we take a worldwide mosquito tour, we will see a few pestiferous species around the globe wherever larval habitats and human blood are available. In addition, we'll see problem species that are tied to a particular region by various ecological requirements, and within their region they have taken advantage of people as a blood source. When mosquito species like to bite various animals as well as humans, there is the risk of transmission of what are essentially animal pathogens (for example, West Nile virus) to people. Let's embark on a mosquito trip and see their quaint customs, taste their bites, and maybe bring home some souvenir diseases.

The first stop is Europe, cradle of Western civilization and a human destination for many millennia. In the north, Scandinavia shares some of the same species as northern Russia, Siberia, Alaska, and northern Canada. It's a short season, but a jolly one. The glacially pock-marked landscape creates many sites that flood with shallow pools in the spring. Eggs that were deposited on the damp soil during the previous summer hatch when they are flooded, the larvae develop as quickly as the temperatures allow, and the adults emerge in hordes. Of course, they bite during the day—and there are over 20 hours of daylight when these suckers are active. This is a case of a really natural mosquito problem not particularly associated with human activity. These mosquitoes have to go about their business quickly before freezing weather ends the season.

Go south a bit to Central Europe, and you will find another mosquito with a similar egg-laying strategy, but this one has as many generations per year as rainfall and temperature will allow. Originally from Africa, this species was carried worldwide, presumably as eggs in soil ballast. They do well wherever human activity has disrupted drainage systems, and they can get up into storm fronts to move hundreds of miles.

Let's move down to the Mediterranean. Here, there is yet another species of mosquito that lays its eggs in soil, but this one tolerates salt water. Combine a lot of shoreline (that is, the Mediterranean Sea) and tides, and you've got a mosquito problem wherever the shore is marshy. Farther inland from the shore are malaria mosquitoes taking advantage of irrigation projects, laying eggs directly on the water at the shallow edges of ditches where vegetation grows.

Enough of the soft life in Europe; let's do some ecotourism in sub-Saharan Africa. True to its promise, tropical Africa has more kinds of mosquitoes and the most severe mosquito-borne diseases in the world. The continent has a couple of species of malaria mosquito that take advantage of irrigation, poor drainage around houses, and urban ditches. Even when these species are not particularly numerous, their affinity for larval habitats near people and their predilection for feeding on human blood make them the worst malaria transmitters in the world. Back in the 1940s, one of them was introduced into Brazil. An intense epidemic of malaria in the Amazon basin followed, which was only stopped by the heroic efforts of Fred Soper, an entomologist who organized an army of employees to completely eradicate the species from Brazil. The mosquito that usually transmits West Nile virus also originates from Africa. Known in the United States as the house mosquito, this species lays its eggs in raft-like groups on the water in a wide variety of habitats ranging from containers to sewage lagoons. They actually prefer to feed on birds, but when they develop in the city they seem to develop a taste for humans as well. The house mosquito is good at getting inside through tiny openings, or it will filter in through open doors during the day and bite at night. That annoying whine in your ear is usually the house mosquito.

The Middle East never seems to be in the news—for its mosquitoes. The problems are locally severe, but where there is no water there are no mosquitoes. Biologically, the region presents an interesting situation in that it has broad connections to Europe, Asia, Africa, and India. At least in the case of malaria, its central location seems to have provided malaria mosquitoes for all occasions throughout the area. Uncontrolled, there have been nasty little outbreaks of the disease associated with oases, containers, irrigation ditches, or marshes. The good old yellow fever mosquito has a toehold on the Arabian peninsula, where it supports transmission of dengue on the shores of the Red Sea—perhaps not so strange, considering that the scientific name of this mosquito is *aegypti*.

India has an impressive problem with mosquitoes, principally its own species of malaria mosquitoes, the yellow fever mosquito, and incredibly high populations of the house mosquito. Basically, it is a matter of many people, numerous habitats, and a warm climate combining to create great conditions for human-adapted mosquitoes. India even has a malaria mosquito that likes to live in the city.

Southeast Asia used to be called Indochina. The mosquito situation is perhaps more like that of India than China, but there is an important difference. The main malaria mosquito of Southeast Asia insists on living under trees so that malaria tends to dissipate when forests are removed. On the other hand, the yellow fever mosquito and the similar Asian tiger mosquito do very well with all the human activity, making dengue fever a severe health threat in the region.

The Australian mosquitoes are distinctive, though none rear their young in pouches. A couple of viruses transmitted by mosquitoes in the temperate and subtropical portions of the country have been the occasion for considerable concern about evening and night-biting species. Farther north, in the tropical areas, we encounter some of the species of the Pacific islands that develop in salt water or near the sea. The yellow fever mosquito occurs there too, and in conjunction with a local species that tolerates hotter, drier conditions, it supports dengue virus transmission. If you go to the oceanic Pacific islands, you'll still find the yellow fever mosquito, but you'll also find related species that develop in small containers like coconut shells. Together, these have been the cause of some extensive epidemics of dengue.

East Asia and Japan make an interesting case for mosquito distribution. They have their own species, which develop in everything from seaside rock holes to rice fields, but they also have some tropical species that migrated as far north as their physiologies allow. A few of these species, including the Asian tiger mosquito, are the only representatives of what are otherwise strictly tropical groups. For the visitor, the ubiquitous rice fields are probably the major contributors to mosquito misery. Not only do the fields go through cycles of flooding and drying during the warm months, but often there is also a big series of ditches and ponds associated with the irrigation.

The Americas are a big destination for the tourist seeking mosquitoes. Particularly in South America, the visitor can enjoy an almost bewildering variety of species. In fact, the variety really is bewildering. Sometimes a single puddle will have eight kinds of larvae or a container will have four or five. There are some stunningly beautiful mosquitoes decorated in iridescent blue, green, or crimson scales. No doubt, people would really appreciate this biodiversity were it not that many, many species are hungry for human blood. Day, night, city, forest, farm—it doesn't matter, there will be a mosquito to greet you. The yellow fever mosquito was eradicated from much of South America during the 1960s and 1970s. The world took its collective eye off the ball and the dedicated work of decades was reversed in a few years. Now, you can have dengue with your *feijoada*, *bife a caballo*, or *sancocho*.

Mosquitoes are generally abundant in North America, though the disease threat is much less than in the tropics. The Pacific coastal states have some of their own species, but they also share a few others with the rest of the continent.

California, Oregon, and Washington have managed to eradicate the yellow fever mosquito, the Asian tiger mosquito, and a third container-breeding species introduced to the United States from Japan in 1998, but the house mosquito is well established. The yellow fever mosquito and the Asian tiger mosquito can be abundant around homes in the southern and eastern portions of the United States. Without them, there would be few mosquito bites during the day in urban areas, but, with them, gardening and working outdoors can be a trial. Floodwater mosquitoes (they lay their eggs on soil in anticipation of flooding) are particularly abundant and varied in North America. They tend to be aggressive biters and occur in outbreaks because they all hatch at once in response to heavy rain or a rising river. The same species that is important from Africa into Central Europe can be a big plague in North America, sometimes moving hundreds of miles in weather systems over the Midwest. Malaria used to be a huge problem in North America. Fortunately, the parasite that causes the disease has been eliminated, but the native mosquitoes that carried the disease remain. Every once in a while, we witness an outbreak of locally transmitted malaria, a reminder of bygone days when Civil War soldiers suffered attack rates as high as 800 malaria cases per 1,000 people per year in Virginia. Finally, a native mosquito of the western states, charmingly named the western encephalitis mosquito, is an important pest that emerges in huge numbers from flooded, grassy pools—a common feature of the area following spring rains or irrigation. This species often occurs in vegetated, decorative ponds in California. You have to wonder if it is domesticating itself to the urban environment—and what other pestiferous species will follow.

After that review of the worldwide mosquito distribution, you would be justified to conclude that there is no region free of them, except for Antarctica and the dry parts of deserts. That doesn't do you much good unless you are planning a trip to the moon. However, within each region, there are some situations where you are more likely to be bitten than others. In urban areas, avoid places with many water-filled containers, neighborhoods where storm drains are stagnant, and locations adjacent to vegetated ornamental ponds, stagnant ditches, or lakes. The same rules apply in suburban locations, but you might also avoid housing near hedges or dense trees, because these may serve as resting places where adult mosquitoes concentrate. If you venture into the countryside, you can stay away (like a mile away) from obvious sources of mosquitoes like vegetated lake shores and marshes. In general, higher, windier sites have fewer mosquitoes: however, some species seem to specialize in sneaking up from the downwind side and biting you in the...

But there is even a finer-scale aspect of mosquito distribution that can be surprisingly helpful in some situations. House mosquitoes most often come from sources near ground level, so occupying a room 10 or more stories up offers considerable protection. The yellow fever mosquito tends to bite low on the body and loves to rest among dark objects so sitting on a raised bed with your feet off the floor prevents many bites—especially if someone nearby is more exposed than you. Sometimes, moving just a few yards outside of a shaded, wooded area and into the sunlight discourages some kinds of mosquitoes from biting (of course, it also puts you in the line of fire for ticks). The presence of animals can also be very helpful. Bites from the house mosquito are unlikely if a coop of chickens is nearby because they usually prefer birds.

BLACK FLIES

Black flies are marvelous examples of insect evolution, but their irritating daytime bites can make some areas difficult for people and animals. They develop in moving water, the larvae clinging to rocks or other objects (sometimes other aquatic insects) on pads of silk they spin beneath themselves. Some people have been surprised in recent years to see an increase in the numbers of these pests in suburban areas where relatively slow streams have become less polluted. Formerly, when pollution and siltation were less severe, some slow-water species coming out of the Mississippi River basin achieved such huge populations that they killed cattle by packing themselves into the animals' lungs.

Black flies occur all around the world, but they are most abundant in portions of the mountainous wet tropics (Central

America, West Africa) and in northern areas where a previously glaciated landscape has created many streams. Black flies are scarce in some areas that otherwise seem like great habitats (for example, Southeast Asia). Within an area where black flies occur, the beasties are likely to be most abundant near flowing water. From a human perspective, the distribution of biters is even spottier because only 10%–20% of species commonly bite people. If populations are low, the threshold of annoyance may be closer to the aquatic sources of those particular human-biting black flies that occur in the area. If populations are high, the ability to fly many miles will result in annoying populations across large areas, independent of the locations of larval sources. Some species of black flies like to bite where hair meets skin—places like the hairline on the head or hairy arms. Others bite low on the legs and ankles. They tend to walk around in an irritating way before biting, which they do by lacerating the flesh, making a mosquito bite seem pleasant by comparison. It's possible to stop a lot of bites by placing susceptible parts of the body where black flies aren't—like inside a hat or long sleeves. In some areas, wearing light-colored clothing discourages biting by black flies. A relatively nice thing about black flies is that they almost never bite indoors or at night.

SAND FLIES

Sand flies are part of a family of flies that includes the commonly observed drain flies. If you open your mind to the possibility, drain flies are kind of cute. They have large wings, they do not infest food nor bite, and they are definitely fuzzy. Sand flies, on the other hand, have narrower wings that stick up, they bite, and their sparse scales stick out—giving them a hard-edged, punk look.

Sand flies came into prominence in the twenty-first century as a source of discomfort and illness for American soldiers serving in Iraq. But sand flies occur in many parts of the world's temperate and tropical regions. With the exception of much of India and Southeast Asia, all of the tropical areas have significant problems. Temperate and subtropical areas are largely free of serious sand fly problems, except for the

Mediterranean region, the Middle East, and northern China. Although there is a lot left to learn about the larval sources of sand flies, they usually seem to be associated with rodent burrows, farming practices, primitive living conditions (including military encampments), and some tropical forests. The adults do not fly very far; therefore, it should be possible to avoid them by avoiding larval development sites. Unfortunately, larval sites are hard to recognize. Most sand fly species bite at night, and many seem to gather on walls indoors before and after feeding. Unfortunately, avoiding walls at night is rather impractical advice for someone trying to avoid sand fly bites.

BITING MIDGES

The U.S. Marines nicknamed these biters "sand fleas" or, more colorfully, "flying teeth." They occur almost everywhere, but the most troublesome human pests are the black gnats that occur in dry lands that have alkaline water sources or along tropical or subtropical beaches. The black gnats can be so abundant and pestiferous that they are sometimes blamed for one of the plagues raised against Egypt by Moses (with a little help). The wise traveler might be well advised to ask pointed questions about gnats before making reservations at a tropical beach resort. If possible, accommodations even a short distance from the shore might help you to avoid long evenings of welts along with your Mai Tais. Some black gnats dislike hot, sunny conditions, and all biting midges dissipate when it is windy. Being indoors helps, especially if you keep the lights out, but biting midges are small enough to penetrate most kinds of window screens.

SNIPE FLIES

It's a mark of distinction to be able to say you were bitten by one of these flies. First, most people don't recognize them, so they'll be impressed by the identification. Second, they usually occur in high mountain places that are visited by the more rugged types of outdoors enthusiasts. They are only a notable problem in local areas ranging from Alaska to Chile, in Australia, and, rarely, in the mountains of Central Asia. They would hardly be worth mentioning, except that snipe flies can be abundant (to the tune of dozens biting at once) and inflict painful bites as they suck blood. Some are more attracted to the head and others aim lower. They bite during the day outdoors. Before you get too excited about your first bite, remember that, in Texas and Mexico, it might have been an athericid, a closely related fly in a different family.

HORSE FLIES AND DEER FLIES

Few could argue that horse flies and deer flies aren't the most beautiful of the biting insects. Sure, they usually lack the iridescent colors of some of the tropical mosquitoes, but they often have fine, Persian carpet patterns over their bodies in a wonderful variety of earth tones. Combine that with a robust, graceful form; reflective red, purple, and green stripes on the eyes; and a splash of crimson where they bite—well, you've got yourself a showy insect. Alas, their beauty usually goes unappreciated, especially where they persistently pursue their hosts, slashing a real wound in the skin and lapping up the resulting flow of blood. The largest species are over an inch long, making you wonder how they ever get a full blood meal without disturbing even the largest of animals.

It's hard to avoid horse flies and deer flies. They occur from north to south on all continents except Antarctica. Generally, there are more species around than the few that bite humans, but those few can be very troublesome. The larvae occur in damp soil, flooded areas, stream banks, and other aquatic or semi-aquatic sites. As a result, the adults are usually most abundant near water, but they often concentrate at the edges of trails, ready to pounce on passersby. Many horse flies and deer flies locate their prey visually. As a result, there is a greater chance of being attacked if you stand out from the natural background. Once you stop moving, the flies that tracked your movement may catch up and bite, but additional flies will have a hard time finding you. Unfortunately, a person outdoors on a trail or along a stream during the day is usually on the way to somewhere else, continuously attracting more horse flies and deer flies.

STABLE FLIES

Stable flies look very much like a house fly, but instead of a soft little spongy thing at the end of their drinking straw, they have a hollow dagger. They are persistent and painful biters, causing annoyance to humans and real economic losses to livestock. Stable flies occur worldwide in temperate and tropical climates. There are similar kinds of flies that only occur in Africa or Asia, but almost anywhere you go, you can find these pests. Stable flies can travel substantial distances if they do not find what they need close to their larval homes. The larval home is putrid vegetation (like compost), manure mixed with vegetation, or manure with a high content of undigested vegetable matter. Before you buy that lovely home next to a cattle feedlot, you might want to think about the stable fly problem. Certain beaches also have severe stable fly infestations, but this may be a result of their numbers accumulating as traveling flies reach the coast and can go no farther. Stable flies only bite during the day, and they seldom come indoors (though even a single one inside can be cause for panicked newspaper smashing). They seem to like to bite on the legs and feet; you will get fewer bites if your guests are barefoot and seated at a lower level.

TSETSE FLIES

The word *tsetse* is supposedly based on the sound that the flies make, but perhaps the most memorable association is with Ogden Nash's short poem "A *Glossina morsitans* bit rich Aunt Betsy / Tsk tsk, tsetse."[1] Purists say that the word *tsetse* implies that these insects are flies, but since most of us require a little background in our vocabulary, we'll stick with tsetse flies.

And we'll hope that tsetse flies don't stick with us. They are incredibly strong fliers, buzzing around at up to 20 miles per hour (32 kph) to catch crocodiles, large mammals, humans, and other animals. Both males and females take blood as their only food, feeding every three to five days on large blood

1. The poem is titled "*Glossina Morsitans*, or, The Tsetse."

meals extracted from their prey through a long, pointed drinking tube. Instead of the usual insect pattern of laying dozens or hundreds of eggs at a time, tsetse flies nurture a single offspring in a womb-like organ. They deposit the large maggot (nearly as big as its mother), where it almost immediately pupates. The good news is that you would have to go back tens of millions of years to find them in the Western Hemisphere; the bad news is that they are abundant in the tropics and subtropics of sub-Saharan Africa. The periodic existence of isolated populations in the Arabian peninsula makes us wonder whether this is yet another insect that might be spread farther to other parts of the world. Within their range, tsetse flies occur in distinct "fly belts" that correspond to favorable habitats for the particular species. In western Africa, this is most likely to be a stream, shore, or forest. In eastern Africa, this is likely to be an open savannah. The flies do not occur above 5,000 feet (1,500 m) elevation, and they do not tolerate even moderately cool temperatures. As a result, you can avoid tsetse flies by staying away from African forests, savannahs, and rivers—a tough thing for tourists who flock to the continent to see forests, savannahs, and rivers. For the most part, tsetse flies bite outdoors during the day and are attracted to large dark objects that breathe.

FLEAS

The symbolism of fleas goes deep, deep into our language and culture. John Donne's famous lines[2] about the intimacy of joining the blood of two people in a single flea seem pretty strange now, but probably made more sense in an era when the wealthy wore elegant flea traps around their necks because of the heavy infestations or when Georges de La Tour painted a picture[3] of a woman killing a flea by candlelight. We still have flea markets, stir our chemistry experiments with magnetic fleas, refer to worn-out things as flea-bitten, and pluck the strings of the ukulele.[4] The very word *flea* implies something

2. "The Flea."
3. *Woman Catching a Flea*, 1638.
4. A Hawaiian word that roughly translates as "jumping flea."

small, quick, and unpleasant. A close look at a flea will reveal a strange and extremely specialized insect. They all bite, most are shaped specially to slip between hairs or feathers, and they are equipped with the most efficient jumping mechanism in all of nature.

Fleas occur everywhere, and, like lice, we have probably taken them to Antarctica at one time or another. Although most species do not bother people, quite a few of them do. Principally, we are bothered by animal fleas, human fleas, and burrowing fleas.

The animal fleas include the cat flea, which actually bites a wide variety of animals. It depends on a close association between the animal hosts and the places where larvae develop, places like pet beds and carpet. Keeping pets out of houses can provide a lot of relief to human occupants, especially in cold climates where the larvae cannot survive the winter outdoors. If an infested house is unoccupied, some fleas have the disturbing ability to hide as mature fleas in their pupal cases, then suddenly spring out when they sense someone enter. This creates a wave of fleas, the attack of which is like something from a science fiction movie. Avoiding buildings with resident dogs or cats is one way to place yourself somewhere these fleas aren't. The same advice applies to rodent-infested cabins, campsites with animal burrows, and other places where animal fleas might like a change of fare from their usual furry prey.

Human fleas also attack a number of animals, but they seem to be able to sustain themselves in homes without pets. Some of these fleas like to be in clothing and bedding, close to where they can get a bite. This habit makes them transmissible, as they catch a ride with laundry, luggage, or rags. You might bring human fleas home following a visit to a dormitory, poorly run hotel, or transient encampment.

Burrowing fleas are nasty pests. People sometimes get sticktight fleas from chickens, but the prime human problem is the chigoe or jigger flea. Their species name is *penetrans*, and do they ever! Occurring in most of the tropics and even parts of the subtropics, the females burrow into the skin (usually on the foot), making a not-so-little cell with her rear end pointing out of the hole. She distends to the size of a small pea and sheds eggs out of the hole. They attack many kinds of animals, and it is not uncommon to see limping people and dogs biting at their toes in chigoe country. The risk from these fleas is much worse in rural areas, probably because its life cycle goes most smoothly on pigs. They also seem to prefer sandy soils, including beaches. It is possible to avoid many bites by avoiding areas where you see affected people and dogs. Shoes help, but a chigoe flea may come home with you and then burrow after you've taken off your footwear.

That takes care of where you can get bitten. In the next chapter, we'll look at some of the reasons that you don't want to get bitten.

WHEN IS ENOUGH ENOUGH?

Malaria, bubonic plague, typhus, dengue, sleeping sickness, dysentery, anthrax, encephalitis—the entire array of arthropod-associated diseases accounts for about 10% of the time lost to illness worldwide. We may not know the whole story behind these diseases, but we know we don't want to get them. The world can be a dangerous place, but, as we saw in the last chapter, the risk is not the same everywhere. By the same token, not every bug bite has the same risk of making you sick (Table 3.1).

Before spending a few pages on a disease-by-disease account of why you want to be more careful in certain places, there are a few general principles that might help in understanding the problem. First, some bites are their own problem. Dozens, hundreds, or even thousands of bites at a time are a possibility following the simultaneous emergence of floodwater mosquitoes, during the season for black gnats, or on entering an unoccupied flea-infested home. Receiving so many bites over a short period of time is more than uncomfortable, even if the event does not result in disease. Of course, some people are more sensitive to bites than others. Those reactions can cross the line into more problems like secondary infections or large-scale allergic responses. The likelihood of such a problem is the result of a complicated relationship among the individual person's susceptibility, his or her history of bites, and the particular kind of bug doing the biting. The bug-associated diseases that get the most attention by scientists, if not by the public, are caused by pathogens carried in the bite. Some of those pathogens get carried on the mouthparts of the insect and are transmitted to someone in the same way that a dirty hypodermic

Table 3.1 Summary of Diseases Caused by Pathogens Transmitted by Arthropods (Vectors)

Disease	Pathogen That Causes the Disease				Arthropod Vector													
	Virus	Bacteria	Protozoa	Roundworm	Mite	Tick	Louse	Bed Bug	Kissing Bug	Mosquito	Black Fly	Sand Fly	Biting Midge	Snipe Fly	Horse Fly	Stable Fly	Tsetse Fly	Flea
African sleeping sickness			×														×	
Anthrax*		×								×					×	×		
Babesiosis			×			×												
Bunyamwera, Bwamba, Ilesha, and Pongola diseases	×																	
California encephalitides: Jamestown Canyon, La Crosse, Snowshoe Hare, Tahyna, others	×									×								
Carrion's disease, oroya fever, verruga		×										×						
Cat scratch fever, bacillary angiomatosis		×																×
Chagas disease			×						×									
Chikungunya, o'nyong-nyong, Ross River disease	×									×								
Colorado tick fever	×					×												
Crimean-Congo hemorrhagic fever**	×					×												
Dengue	×									×								
Eastern equine encephalitis	×									×								
Ehrlichiosis		×				×												
Endemic typhus		×																×
Epidemic polyarthritis disease	×									×								
Epidemic typhus		×					×											
Filariasis, elephantiasis				×						×								
Hemorrhagic fever with renal syndrome*	×				×													
Japanese encephalitis	×									×								
Leishmaniasis: Kala azar, espundia, Baghdad boil			×									×						
Loiasis: Calabar swelling, eye worm				×											×			
Lyme disease		×				×												
Malaria			×							×								
Murray Valley encephalitis, Australian X disease	×									×								
Onchocerciasis, river blindness, craw craw				×							×							
Oropouche fever	×												×					

Disease	Pathogen That Causes the Disease				Arthropod Vector													
	Virus	Bacteria	Protozoa	Roundworm	Mite	Tick	Louse	Bed Bug	Kissing Bug	Mosquito	Black Fly	Sand Fly	Biting Midge	Snipe Fly	Horse Fly	Stable Fly	Tsetse Fly	Flea
Q fever**		×				×												
Relapsing fever		×				×	×											
Rickettsial pox		×			×													
Rift Valley fever**	×									×								
Rocky Mountain spotted fever, Boutonneuse fever, many other spotted fevers		×				×												
Russian spring-summer encephalitis, tick-borne encephalitis, Kyanasur forest disease, Omsk hemorrhagic fever	×					×												
St. Louis encephalitis	×									×								
Sand-fly fever, phlebotomus fever	×											×						
Scrub typhus		×			×													
Sindbis, Ockelbo, Pogosta disease, Karelian fever	×									×								
Tularemia, rabbit fever**		×				×				×					×			
Venezuelan equine encephalitis	×									×								
Western equine encephalitis	×									×								
West Nile fever and neuroinvasive disease	×									×								
Yellow fever	×									×								

*Arthropods are a minor part of transmission.
**Infection from direct contact is also important.

needle spreads infection. Most of the bug-borne diseases are exquisitely adjusted to the relationship between the biter and the bitten. The pathogen must become more numerous in the bug, sometimes changing through various forms. This incubation period can require days or weeks, the speed usually dependent on temperature. Some of the organisms causing disease typically infect animals; human infection is a relatively rare side effect for those diseases called "zoonotic" by epidemiologists. Other organisms exist almost exclusively in humans and in particular biting bugs. The animal diseases are often rarer in humans, but it becomes almost impossible to eliminate them from an environment where their animal hosts are constantly exposed to bugs. The complete elimination of diseases that only affect humans is a real possibility because the target organism is accessible in humans as well as in the bugs that carry it. On the other hand, human diseases are most common where there are, well, more humans—not exactly a recipe for good public health.

We have already identified various kinds of biting insects, ticks, and mites in chapter 1. Let's go through them again, considering the risks from their bites.

MITES AND CHIGGERS

Mite bites are often amazingly irritating, considering the tiny size of the biters. Chiggers can cause an irritation that lasts for weeks, sometimes creating a fiery ring of itchiness around the belt line. Rat mites can drive you crazy when they are numerous, each one leaving a point of irritation. As with most bug bites, some people develop a hypersensitive reaction following many bites. As unpleasant as these problems are, some mites create a much more serious problem by transmitting disease-causing bacteria. Rickettsial pox carried by mouse mites is no fun, but seldom serious. Scrub typhus, on the other hand, can be a truly unpleasant disease, and it is sometimes deadly. What's more, the scrub typhus infection can be very common in some populations of Asian chigger mites. If you are hiking, farming, or fighting in areas with scrub typhus, even one chigger bite may be too many.

TICKS

The large variety of ticks makes it hard to generalize about them. Most people who are active outdoors have been bitten by a tick at one time or another. Although the skin may react to the bite with a reddened area, or a secondary infection may extend the problem a few days, the bite itself is usually not a big deal. However, some hard ticks and not a few soft ticks have a bite that is very painful. The pajaroello soft tick in the western United States and Mexico has been described as having one of the most painful bites of any arthropod, though experts claim that such a painful reaction is truly exceptional. The right (or wrong) hard tick bite above the spine can cause alarming, but reversible, paralysis. The real fear from tick bites comes from the multitude of diseases caused by organisms growing in the ticks. There are viral diseases (Colorado tick fever, Congo-Crimean hemorrhagic fever, and tick-borne encephalitis), bacterial diseases (ehrlichiosis, Lyme disease, Q fever, relapsing fever, Rocky Mountain spotted fever, other spotted fevers, and tularemia), and a protozoal disease (babesiosis). Many of these are not lightweights: they can sicken, mutilate, and kill. The wrong tick bite is definitely one too many, so it is important to know the situation in a local area if you might be exposed to ticks. Often, the simple solution is to get the tick off promptly (before eight hours of attachment), but it is even better to take precautions to prevent the bite in the first place.

LICE

Ewww! Ick! Lice. Cooties. Nobody wants even one louse, but they seldom operate alone. All too often, the infestation isn't noticed until the hair is full of nits (eggs) or until itching in intimate spaces drive a person to distraction. Persistent infestation leads to some typical syndromes, like pallid soft skin, crusty irritations, and stained fingernails (stained with crushed lice—ewww! ick!). At least, a lousy person is never alone. Of the three kinds of lice that infest people, only the body louse is known to transmit disease organisms. They are all bacteria, and they cause one of the most devastating epidemics known (typhus) and a couple of miserable diseases (trench fever and relapsing fever). Human lice only infest humans, therefore, it is possible to eliminate them from entire communities. Once lice get started, they spread rapidly, and the consequences can be serious indeed.

BED BUGS

Bed bugs are sometimes unbelievably numerous, sucking significant quantities of blood from a person exposed to the infestation. The bites range in severity from little more than a rash, to welts similar to those caused by a mosquito, to large fluid-filled blisters. Occasionally, a person will become highly allergic to the bites and have a bad reaction that causes a real illness. Considering the unpleasant nature of an infestation—dozens of bugs sneaking out of their hiding places at night to fill up on the blood of a sleeping person—it is not surprising that sleeplessness is a common symptom. Remarkably, not one disease organism has yet been proven to be transmitted by common bed bugs. This is especially surprising because it seems that they would be the perfect vector, biting repeatedly and living a relatively long time. Currently, people who don't have bed bugs call them a nuisance; those who have bed bugs call them a disaster.

KISSING BUGS

One of the Native American languages of South America distinguishes kissing bugs as the "big bugs." They are commonly 1" or 2" long and swell up to great size when they feed. The bite is usually painless, and the bug is sometimes careful not to step on the skin so that a person is seldom awakened by these vampires. On the other hand, certain people develop a severe allergic response to the bite of some of the North American species.

Kissing bugs commonly carry a serious protozoan pathogen in their guts, and the resulting infection is called Chagas disease. Woe to him (or her) who acquires this illness. Often, the parasites lodge in the heart and cause permanent damage years after the bite. The treatment for Chagas disease has never been perfected, making a single bug's "kiss" potentially fatal. Well, strictly speaking, it is not the bite that transmits the disease organism. A person is infected from the feces excreted by the bug at the time of the bite.

MOSQUITOES

It is a rare person who has never been bitten by a mosquito. Who expects such a mundane event to be fraught with danger? In some parts of the world, a single bite can lead to illness and death. There are a number of reasons that mosquitoes are so good at biting and at spreading pathogens that cause disease. First, the physical attributes of most mosquitoes are adapted to blood feeding. From the long legs for posture to the sensitive antennae and from the highly adapted blood-digesting gut to the intricately constructed feeding tube, a mosquito is a host-seeking, blood-sucking, digesting machine. Second, mosquitoes come in thousands of forms and thus vary in their strengths and predilections. As a result, from a pathogen's point of view, there is likely to be a mosquito somewhere or other that satisfies its requirements for transmission. Finally, mosquitoes usually bite repeatedly during their lifetimes, giving them the opportunity to pick up a pathogen, let it multiply in their bodies, and then inject it into another host—maybe you.

Mosquitoes transmit a great variety of viruses, a few protozoa, and a few roundworms. Mosquitoes occasionally transmit the bacteria that cause anthrax and tularemia, though the transmission is through simple contamination of the mouthparts. Let's start with the worst pathogens, or maybe the best—it all depends on how you look at it. For example, perform a little word association with "snowshoe hare" or "Jamestown Canyon" or "La Crosse." You get something warm and fuzzy, a likely scenic destination, and, perhaps, cheeseheads. In fact, these are all names of viruses that occasionally infect people but that rarely cause disease. Even the most prevalent and serious one, La Crosse, occurs rarely, and a person would be labeled careful indeed if she spent a lifetime avoiding exposure to this virus.

There are other viruses that are also pretty rare, but they sometimes appear in outbreaks where the local risk of transmission is quite high. Some of these are serious illnesses that can cause permanent damage or even death. The list is extensive, but perhaps Chikungunya, Japanese encephalitis, Rift Valley fever, St. Louis encephalitis, or equine encephalitis (choose from Western, Eastern, or Venezuelan) ring a bell. If you receive warnings of an outbreak of any of these viruses in your area, it would be well worth your while to avoid all mosquito bites.

A few viruses transmitted by mosquitoes occur over large areas, though they might be more common at one time or another. These include two of the grim reapers of medicine: dengue and yellow fever. Our old friend, the yellow fever mosquito, is the principal culprit for the transmission of both viruses. We already know how easily this mosquito gets along with people all around the world wherever the average winter temperature stays above about 50°F (10°C). Dengue occurs in four types, each of which can infect a person once in a lifetime. Children often have no symptoms at all, but a few kids develop a rapidly fatal illness, especially if they have been infected a second or third time. Adults get sick more often than do children, but they seldom die. Some adults get so sick that they wish they were dead. Typical problems are high fever, extreme pain in bones and nerve trunks, and a long depressing period of recovery. In dengue country, don't get bitten, especially by mosquitoes that bite low on the body during the day. Yellow fever used to occur in huge outbreaks, sometimes depopulating whole towns. Yes, depopulated, as in so many dead that the town folded up. Yellow fever was behind the whole Walter Reed story that eventually freed Panama from this scourge so that workers could live long enough to dig the canal. The discovery of a good vaccine for yellow fever really yanked the teeth of the disease, but there are occasionally large outbreaks. Don't go to yellow fever territory without vaccination.

One bite from an infected mosquito is usually enough to cause a viral disease. In contrast, it takes many roundworms to make a person sick, and only a few are transmitted in each bite. The worm breaks through the sucking tube's sheath to enter a person's skin through the hole drilled by the mosquito. A variety of mosquitoes transmit roundworms, but the tropical house mosquito is the main one. These days, filariasis (the disease caused by the roundworms) is restricted to tropical areas, but that still means that 1.2 billion people are at risk. In the good old days, it was common in many more tropical and subtropical cities. Now the extent of the disease is shrinking fast thanks to a global effort that relies on drug treatment of people and mosquito control. You are not likely to get filariasis from a few infected bites, but a lot of exposure could result in fever, itching, and, eventually, gross swelling of the legs and groin.

Malaria is one of the big five diseases transmitted by arthropods. We've mentioned typhus (transmitted by lice), filariasis, and yellow fever, and we will talk about plague when we get to fleas. Of these five terrible diseases, only malaria remains a daily, tragic problem for hundreds of millions of the world's population (Figure 3.1). The parasites go through a complicated life cycle in humans and in one type of mosquito (*Anopheles*), their only hosts (Figure 3.2). Malaria comes in four forms that cause two kinds of disease. One form is very severe and often fatal if not treated. The other is milder, seldom fatal, and may hang around in the body for years. Symptoms vary tremendously, but the chills, pain, nausea, and severe headache are seldom described as mild by those who suffer. Tragically, people who get sick after returning from a malarious area may not get prompt treatment because the early symptoms resemble influenza and other common diseases. Effective drugs usually treat malaria successfully if they are given in time to prevent death. The malaria protozoans have developed resistance to many drugs, but it is still possible to take pills that prevent infection in most areas. Sometimes, people do not want to bother with the expense, inconvenience, and side effects of preventive drugs because they know they will only spend a few days in the area. Under those circumstances, avoiding mosquito bites becomes especially important. A single bite from the wrong mosquito can cause a memorable illness or even death. It happens every day many thousands of times.

BLACK FLIES

Black fly bites affect some individuals very badly. Of course, when black flies are numerous, they would drive anyone crazy. Certain people get a horrible reaction to even a single bite, with swelling, pain, and even scarring. A syndrome called black fly fever occurs when enough bites cause headache, nausea, and fever. In Central America, western Africa, and southern Yemen, black flies transmit a roundworm that causes a disease known as river blindness. River blindness strikes people who have had many inoculations of the roundworm, just as the worst symptoms of filariasis affect people who have had many inoculations from mosquitoes. As the number of worms builds up in an individual, they produce many immature worms concentrated in the skin. Itching and thickening of the skin ensue and, eventually, the worms invade the eye, causing blindness. Heart-rending pictures of old, blind people leading each other along village paths illustrated how common the disease was for populations exposed to the flies their entire lives. A large-scale black-fly control program combined with new, effective drugs has greatly reduced the problem.

Figure 3.1. Countries with and without malaria transmission by mosquitoes. Courtesy of the Division of Vector Borne Infectious Diseases, Centers for Disease Control and Prevention.

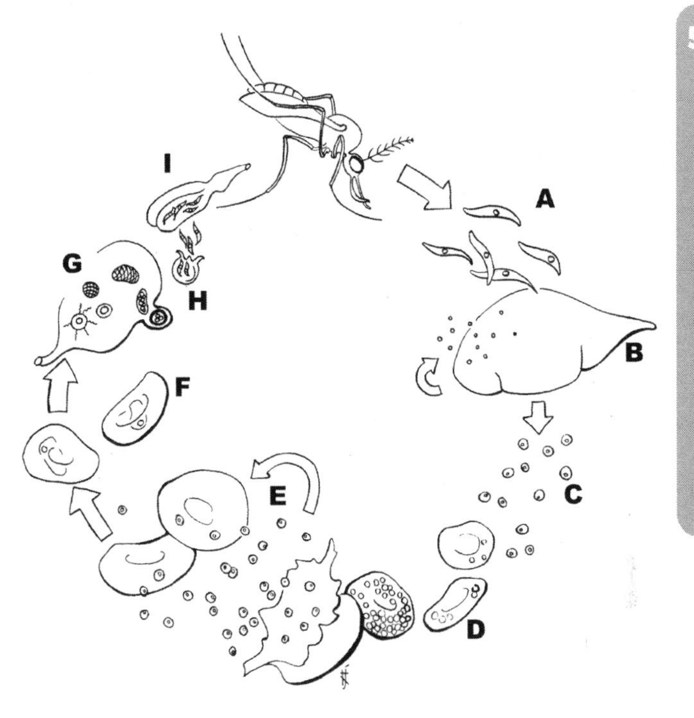

Figure 3.2. The life cycle of a malaria parasite. Starting at the top, the sporozoite stage of the parasite is injected in the saliva when the mosquito bites and then this stage makes a beeline for the liver. The sporozoites (A) infect liver cells (B), multiply vastly in numbers, and escape into the bloodstream as the merozoite stage. Merozoites (C) get into red blood cells (D) and multiply again, infecting even more red blood cells (E). After a while, some infected red blood cells (F) produce malaria sex cells (micro- and macrogametocytes), which are picked up by another malaria mosquito. The mosquito midgut (G) is the love nest for the gametocytes; they join there to form a mobile egg that lodges in the midgut wall. Even more malaria parasites are produced as the egg matures and produces a cyst (the oocyst, H) that generates sporozoites. These burst through the midgut into the interior of the mosquito, migrate to the front of the bug, and infect the salivary glands (I). At that point, the mosquito is ready to infect another person.

SAND FLIES

Roughing it during an adventure tour of Southwest Asia, you've just completed a restless night's sleep on a thin mattress in a large room with many other beds. Adjacent to the dormitory is a single shower consisting of a slimy cement stall with a single pipe projecting from the wall. Any feeling of refreshment from the previous evening's shower has long since worn off during a sweaty night. The only real rest you've gotten was in the few hours before dawn when the temperature descended from unbearable to merely hot. Bleary-eyed, you focus above your head on the chalky wall, which is painted a poisonous shade of blue-green. There, staring back at you, are small flies with their wings held above their backs. They have beady-eyed expressions, and their tummies are full of your blood.

The blood they took was not the most horrible thing that happened that night, however; the most horrible thing was the parasites that the sand flies left behind. The parasites already reproducing in your skin are embarking on their own adventure through your body and will end in your liver and spleen. Untreated, the disease, known as kala azar, will progress to the point that your organs swell, you are prostrate with fever, and you eventually die. The only available treatment is an old drug made from a heavy metal that often has its own side effects. Other forms of the parasite, *Leishmania*, cause ulcers that leave interesting scars (Baghdad boil) or eat away cartilage tissue in the face (espundia). Such diseases occur in tropical America and in subtropical and tropical parts of Africa, Europe, and Asia. Sand flies also transmit a virus that causes the disease labeled, appropriately enough, sand fly fever.

Sand flies seem to have particularly potent salivary secretions that can produce a severe reaction in some people. The bites have an irregular outline and rather uniform coloration and are only a little raised from the skin. The irritation from a bite may last for several weeks. Under the "right" circumstances, a person might get hundreds of bites, creating what appears to be almost a solid rash. Sometimes, people develop an allergic response to the bites, which causes the lips and eyelids to swell. Anyone who has had that experience would be highly motivated to avoid sand fly bites in the future.

BITING MIDGES

Even smaller than sand flies, biting midges can pack quite a wallop with their bites. Some species are so small that people literally do not see them biting. They think they have some unusual skin disorder when they notice the generally round, flattened bites. These black gnats are truly tiny and occur in clouds of biting flies that emerge from the upper layers of the soil. When they are numerous, they give the impression that the air itself is pricking your skin. After a few hours of exposure to black gnats, some people have areas of inflamed skin, swollen glands, and fever. Though the bites are miserable, midges transmit few pathogens. About the most serious is the Oropouche virus in South America, which occurs in rare but nasty outbreaks of febrile disease.

SNIPE FLIES, HORSE FLIES, AND DEER FLIES

These large flies have similar mouthparts and biting habits. Some species are very persistent so that even one fly can annoy. The biting mechanism is like a pair of scissors, and your skin is the paper. Not surprisingly, these bites hurt. Their blood meals seem to succeed more often when there are many flies because it becomes impossible to chase all of them away at the same time. Under those conditions, it's easy to see why cows and horses have long tails that serve as fly swatters. Some of these flies occasionally transmit the bacteria that cause anthrax and tularemia, but the only specific association is between certain deer flies and a roundworm with the euphonious name *Loa loa*. This roundworm likes to head for the eye, generating more gruesome pictures for medical textbooks.

STABLE FLIES

Stable flies may sometimes carry anthrax bacteria mechanically, but their real threat comes from the bite itself. They like to bite on the legs, causing a sharp pain and an itchy welt. It seems like stable flies never give up once they start looking for blood, and they are adept at avoiding a slap. A few stable flies can ruin an outdoor gathering as they continue to attempt to feed on people who take exception to being pincushions.

TSETSE FLIES

Tsetse flies are weird and amazing flies. In addition to being the fastest of the biting insects, they are also the most motherly. Females produce one baby fly at a time in an organ that resembles a mammalian uterus. A tsetse fly's bite is painful and leaves a welt, but it is nothing special compared to such toxic horrors as black gnats and black flies. The bad thing about tsetse fly bites are the parasites they leave behind. An infection with these typanosomes causes the infamous sleeping sickness. In eastern Africa, the disease is rapidly fatal following severe inflammation of the heart and nervous system. In western Africa, it progresses for months or years through various states of debility, starting with swollen glands and rash, coming toward a conclusion with a semiconscious patient, and ending with a dead one. Although there is a drug to treat the illness, sleeping sickness is definitely one of those souvenirs of the tropics to be avoided at all costs. The disease has made a big difference in the history of Africa (750,000 people died between 1896 and 1906), keeping the human population low in whole regions or, in its animal forms, making the husbandry of domestic animals impossible. One of the triumphs of public health has been the control of sleeping sickness by effective control of tsetse flies. These programs require constant attention, however, and fall into ineffectiveness when political upheaval disrupts local governments' functioning.

FLEAS

One of the authors has a vivid memory of walking through his childhood with flea bites on his legs from infestations of human fleas. Later, in Texas, his kids got flea bites from the animal fleas on the family's dogs. These days, people control household fleas much more reliably with new products. A few flea bites are generally not considered a major problem. Some individuals develop an allergic response that creates large welts. Dermatologists scare us with names for the bites, like *purpura pulicosa*, but household flea bites usually dissipate after a few days of itching.

The burrowing fleas are a different story. Unaware, a person wonders what happened to create a little, sore swelling. Then the swelling gets bigger and develops a distinctive, callous-edged hole. Things get more and more painful, especially if, as frequently happens, the flea burrow becomes infected.

People who live with chigoe fleas know how to sterilize the site, enlarge the tiny hole in the skin, and withdraw the flea on the point of a needle.

Certain kinds of fleas transmit the bacteria responsible for plague (a.k.a. the Black Death). These fleas pick up the bacteria from animals, most typically rats, and then transmit the plague bacillus to people. Plague usually also kills the flea, as its foregut gets so stuffed with bacteria that it cannot accept a real blood meal. Of course, plague has been a cruel sculptor of demographics in historical times, but even recently (for example, Zambia in 2001, India and Malawi in 2002, Algeria in 2003, Democratic Republic of Congo in 2006) it has occurred in smaller, significant outbreaks. A single infected bite can transmit the bacteria to an unlucky recipient. The consequences of plague are seldom mild, and often tragic. The bacteria are susceptible to antibiotics, but the disease is usually fatal if not treated promptly.

Fleas and plague go together like death and taxes, but fleas transmit other bacteria as well. They are sometimes associated with such lovelies as tularemia, salmonella, Q fever, cat scratch fever, and staphylococcus. More commonly, they transmit the bacteria that cause endemic typhus and a closely related disease. Endemic typhus causes some rashes, fever, and, rarely, complications of internal organ function. The disease occurs around the world in tropical and subtropical areas, usually cycling between rats and fleas, or transmitted from mother flea to baby flea. The bacteria multiply in the flea to the point that they pass out of the gut through the feces. Transmission is not usually by bite, but by rubbing the flea feces into the tiny puncture created by a bite. The disease used to be a significant problem in some of the cities of the American South, and it may be a lot more common than we think even in the twenty-first century. One of the early missions of the U.S. Public Health Service was to control transmission by using DDT to dust areas frequented by rats. The organization went around the community in "typhus wagons," which were equipped to handle the job. Endemic typhus is generally a mild disease, but there is good evidence that it can sometimes cause severe illness in children.

So, when is enough enough? As usual, the infuriating answer is, "It depends." A single bite infected with the pathogens that cause malaria, plague, epidemic typhus, or yellow fever may be a death sentence. In an area where any of these diseases might be present, it makes sense to take extreme measures to prevent infections and bites. On the other hand, you might feel a little more willing to risk at least some bites if the disease is not usually fatal (for example, dengue in adults) or is exceedingly rare (for example, St. Louis encephalitis).

The energy and expense you invest in avoiding bites can be gauged against the risk from a particular arthropod-borne disease. To start, let's consider situations where a person might reasonably take no precautions at all. First, if the disease is not present where you are, then there is really no need to take precautions. It is easy to think of many examples in places like Canada and Britain because arthropod-borne diseases are blessedly rare in those locations, especially during cool seasons. Even a classic winter threat, epidemic typhus, has been essentially eliminated from those countries. Second, if the disease is present but rare, the consequences of getting the disease are likely to be mild, and personal activities do not favor getting bites, then a person is probably justified in taking no precautions.

The individual usually decides whether or not to use insect repellents and other forms of personal protection from arthropods. That approach means that a person will get at least a few bites before actually taking measures to stop them. This free-choice method will work in situations where many bites need to occur before there is a high risk of getting the disease and where the consequences of the disease are not likely to be permanent. In practice, most people probably use this method even when the potential consequences of disease are severe as long as they perceive that their own chances of getting an infected bite are low.

Most people will not reapply a repellent or take other precautions until they start getting bites. This could be a big mistake in areas where the diseases are severe and there is even a small chance of exposure. If you know there is a severe arthropod-borne disease in your area and that there is even the smallest chance of being exposed, it would be prudent to take protective measures whether or not you see the biting insects. This includes preparing yourself with the right vaccinations and preventive drugs. The consequences of infection with malaria, plague, or yellow fever are simply too great to justify taking any chances.

What about the person who goes to an area knowing that severe disease is present and that she will be exposed to the insects that transmit the pathogens? This is a dangerous situation not to be taken lightly. Every possible measure should be taken to prevent the infection, including devout dedication to getting vaccinated, taking preventive drugs regularly, and preventing all bites. It is the kind of situation that a person who cannot take all the precautions—due to pregnancy or individual sensitivity, for example—should avoid.

As for damage from the bites themselves, an individual's response to a particular kind of bite makes a big difference to the numbers of bites that are tolerable. Most people don't want even a single chigoe flea bite, but many accept a few summer chiggers and mosquitoes as normal. We all know people who react badly to bug bites. Those folks may want to pay special attention to the next few chapters.

STOP THEM AT THE SOURCE

If you currently suffer from a lot of bites, the very concept of a biteless existence seems like an impossible dream. In many situations, it is actually possible to change the local environment so that biting arthropods simply can't get to you. As long as the biter is not one that travels long distances, taking care of things in your own home will solve the problem. Think of the home as a restaurant with you and your family among the menu items. Many kinds of arthropods come in for a bite, but you would like to refuse service. You actually have control over most of the kinds of arthropod clientele described in this book. Biting mites, ticks, lice, bed bugs, kissing bugs, certain types of mosquitoes, some stable flies, and fleas can all be eliminated from the Bug Bites Café you call home by paying attention to some important details (Figure 4.1).

There are some general principles involved in altering your home environment to exclude biting insects and ticks. First, the source of the bugs must be accessible. Although a truly inaccessible site is rare in a home, a couple of examples are an accumulation of water with mosquitoes hidden at the bottom of a long pipe, or a street cat's lair infested with fleas in a very narrow crawl space. The second general principle for do-it-yourself elimination of a source of bugs is that it should be manageable in size. Draining a low area of the yard that holds water in the summer is easy to say, but actually accomplishing the task may involve some major digging, installation of drain tiles, and other actions that are beyond the capabilities of most homeowners.

Figure 4.1. The house gone wrong. From top to bottom and side to side, this little place provides many sources of bugs that will feed on the blood of the family inside.

Most of us have had enough experience with insects to know that there are more of them than there are of us. When bugs are a bother, it sometimes seems that we can only reduce their numbers or pitifully exact revenge on a few of them. Suppose that you are able to eliminate the sources of mosquitoes in your own backyard, but your neighbors have done nothing. Perhaps the number of bites has been reduced in half, but you still have a problem. It is easy to forget that you now have only half the problem and any other measures you take to control the mosquitoes will be more effective. Sure, you would be better off eliminating all the sources of mosquitoes—and sometimes the complete elimination of biting bugs is possible—but partial success is a good, firm step on the road to ridding your home of these pests.

We are going to visit a house of horrors infested with every biter whose favorite saying is "Su casa es mi casa." We will see what can be done to eliminate the bug nurseries from a home layer by layer and inch by inch. Some of the actions are easy and others are really difficult—but bites can be a big motivator for home improvement.

The chimney leads to a home's hearth, the traditional symbol of the center of a family. The chimney also leads directly into the living spaces with only a loosely sealed flu forming a barrier. Some kinds of birds, like house sparrows, swallows, and starlings, may construct nests near the tops of chimneys. Most people would not think of the birds themselves as a problem, though their droppings create a different kind of threat to health. The trouble is that the birds' nests support their own little zoo of bugs, some of which wander into your house and bite. The worst culprit is the fowl mite, which may come into the house in large numbers, especially after the birds abandon their nests. The solution is, of course, to remove the source of mites by removing the nests. Those birds can find a better place.

Now let's get our minds in the gutter—the roof gutter. Homeowners are used to hearing lectures on the importance of gutters to the integrity of a house. But let's face it, if it was easy to keep gutters working properly, there wouldn't be a need for all the lectures. The fact is that leaves and other detritus often block gutters, trapping water. Even when gutters are clean, they can be bent slightly by falling branches, misplaced ladders, or even the misplaced foot of a homeowner who pulled back from the brink in the nick of time. Poorly installed gutters may trap water in portions that don't drain toward the downspout. Regardless of the cause, even a thin film of water mixed with organic debris can make a perfect place for the development of some kinds of mosquitoes. If much of the length of the gutter has water, it may produce hundreds of mosquitoes every time the water level changes, hatching eggs previously laid on damp surfaces. In some climates, condensation on the roof produces enough moisture to creep down the roof and hatch eggs in the gutter, even in the absence of rain. The solution is to keep the gutters squeaky clean and in good repair. Clean gutters will dry out quickly (mosquitoes generally require water for 5–10 days to complete development) and also do their job of protecting the structure of the house. Gutter guards are designed to keep leaves out, so they probably help to prevent infestations of mosquito larvae. If you use gutter guards, it is important to be sure the gutters slope toward the downspout because you won't be able to inspect them easily. (Those lectures just never stop.)

We don't have to go much lower to find another source of biting bugs in the home. If there is a tile roof, the ends of the rows of tiles form hollows that lead up under the entire roof. When these hollows are open, they make a perfect place for birds, bats, and rats. Thatch roofs in the American tropics are generally constructed with no barrier between the thatch and the living space. Thatch is a favorite home for kissing bugs, where they support themselves by emerging at night to feed on humans, pets, and rodents. If the roof of your home is sublet to pests, eliminating the source is not a practical option. You'd better turn to chapter 6, "Good Fences Makes Good Neighbors."

Just under the roof is the soffit, a great word because it proves that entomologists aren't the only people fond of jargon. If properly constructed, the soffit is pierced by a series of vents to allow air circulation in the attic. Those openings usually have louvers or screens over them. Unfortunately, those louvers or screens also often have gaping holes. The holes may be the result of the wear and tear of time, but rats and squirrels have been known to make their own holes. The rodents bring in nest material and make nurseries for biting mites and fleas. Some areas of the United States have a kind of rat that normally makes big nests of sticks outdoors, but they sometimes move into the attics of houses. The nests of these rats support not only mites, but also kissing bugs. Creatures in attics are obviously a problem, but it is surprising how often they go unnoticed until there is a biting mite problem. Watch for big holes in soffits and the dirty streaks rats leave as they brush against surfaces. Bats and squirrels are more obvious, particularly as they come and go from openings to the attic. Sometimes you may need professional help to clear your attic of creatures, whether birds, bats, squirrels, or rodents. It may be impossible to seal the entrances and exits while rodents are away, since they live in those spaces 24 hours per day; however, the last thing you want is to seal up squirrels, birds, or bats, leaving them to a horrific and noisy death in your attic.

The inside of a house is the human nest, infested with people, pets, mice, and rats. All of these animals leave their characteristic residues, such as crumbs, grease, hair, skin flecks, droppings, etc. A home that is never cleaned becomes intolerably dirty in a hurry. Some of the components of household dirt support flea development, but for the most part, the act of cleaning itself, rather than the removal of dirt, discourages pest development. Large appliances, furniture, and the backs of cabinets provide quiet places that are ideal for mice, rats, and their attendant mites. Every time one of those sites is disturbed by cleaning, it makes life a little harder for the pests. Bed bugs also seem to be associated with lackadaisical cleaning. The household dust and dirt does not seem to support bed bugs in any way, but the disturbance and removal of their hiding places discourage the bugs. Untidy habits in the bedroom create many crevices that become bed bug hiding places, so Mother was right: clean up your room!

The proper storage and handling of food helps to prevent rodent infestations. A refrigerator is generally rodent-proof, but almost no other storage container will completely stop them. Thicker plastic containers at least slow them down, but if you have a bad problem, only metal containers will do. Some food sources are not so apparent, like pet dishes, grease under a cooktop, and the accumulation of crumbs in out-of-the-way places. Rodents can infest homes where there are no obvious sources of food, but they certainly do better where food is abundant.

Cats and dogs are the most common cause of fleas in homes. The larval fleas eat material excreted by the adult fleas, creating a flea nursery wherever the pet tends to spend time. Significant sources of larval fleas can be eliminated by washing the covers of pet beds, cushions, and rugs that are the habitual resting place of a cat or dog. Vacuuming removes much of the larval flea food, as well as some of the fleas. Old and infirm dogs sometimes get an infestation of a special tick, the brown dog tick, that can complete its life cycle indoors. Each female tick lays something like 5,000 eggs, so it isn't long before an infested home is crawling with them. The female ticks usually crawl into a crevice somewhere to lay their eggs and die, sometimes creating a disgusting accumulation of tick gunk behind dressers, under drawers, or in other hiding places in the house. This kind of tick rarely bites people, but with 5,000 eggs per female, there will be exceptional individuals that switch from dog to human. Dogs also occasionally bring in outdoor ticks on their fur. The tick may fall off the dog before feeding and then find a member of the human family for its next meal. Whether the house is infested with fleas or ticks, 9 times out of 10, it is the pets that are responsible. You could remove the source by eliminating the pet, but that wouldn't be much fun. You can also cure the pet with any of several products we'll discuss in the next chapter.

Remember the yellow fever mosquito? It loves to live with people: it loves the water that people put in vessels, it loves the blood that people contain, and it loves to be indoors with its favorite companions. In tropical locations where running water is either absent or unreliable, families often store water in jugs, drums, jars, or cement basins. People are storing water for a reason, so they tend to put these containers where they need them, especially in kitchens and bathrooms. Simply emptying a vessel is generally not sufficient to remove all the larvae, because they are very good at diving down to the last bit of water remaining. Adult mosquitoes emerging from these sites have an easy life with humans for blood and abundant locations in which to lay their eggs. Even homes with modern construction and good running water can have infestations of yellow fever mosquitoes in waste water that accumulates in the evaporator pans of refrigerators, under the spigots of water coolers, in pans that catch a drip, and in uneven bathroom surfaces that never drain. When one of the authors lived in Texas at a time when this mosquito was abundant, it was common to find its eggs laid in the rough surface left by hard water on the inside of the tooth mug. Even a small larval site that only produces a few mosquitoes per day is a big problem indoors because practically every female mosquito will bite one of the inhabitants. Many of the small sites are easily eliminated by discarding them. Large sites are more difficult because they usually serve an important household purpose. Emptying large containers can be an energetic household chore and requires a good scrubbing inside to remove residual larvae and eggs. If you live in the tropics and need your containers, you may want to rely on some of the methods described later in this book.

You can pick up lice almost anywhere that poor hygiene is sold, but many infestations start right in the house among family members. It is a little unfair to say that head lice are only the result of poor hygiene, unless you consider the exchange of head gear or a tête-à-tête to be poor hygiene. Nonetheless, when your child brings head lice home from school, it isn't long before the whole family is infested. It is small wonder that infested children are usually barred from the classroom until they are certified louse-free. This is a case where stopping the bug at its source makes a lot of sense. Body lice can get started with any exchange of clothing or close contact, but they are a rare problem in modern homes. Keeping healthy seems to discourage body lice, at least at a population level. Finally, crab lice are restricted to the thicker hairs of the eyebrows, beard, armpits, pubis, and around the anus. Best not to invite a crab-louse-infested person into your home.

How low can these biting bugs go? Why, the basement, of course. Rats and mice are often a problem here, as they were in the attic and in the living spaces. Their biting mites can easily crawl up through cracks or holes in the floor to bite the inhabitants. Those bites often occur in one place, like on one side of a bed or at a particular desk. Considering their tiny size, the mites move big distances, but they probably get channeled by structural barriers to specific locations in the house. Crawl spaces and basements are also subject to flooding, either because of poor grading around the house (and those darn gutters) or because of a leaking pipe. As long as even a fraction of an inch of water remains for a week or two, there is the potential for mosquitoes to move in. In this case, the culprit is usually the house mosquito. They may usually prefer to bite birds, but with thousands of them bubbling up from the basement, some human blood is bound to be shed. Sometimes, mosquito bites and the presence of male mosquitoes in the house are the first signs of a major plumbing leak. With or without mosquitoes, a basement full of gray water is something you want to fix.

A porch is just plain nice. The normal rules of housekeeping don't apply as stringently, cool breezes wash over it in the summer, and it is the first relief from the weather as you enter your house. Take a look under a porch, deck, or elevated entryway, and you may find the same wonderful qualities—here enjoyed by an animal. A wide variety of animals find such spaces ideal for their homes and families. Sometimes, they find their way unobserved into the basement or crawl space through broken vents under a porch. The biting arthropod problems are the same as those already discussed: fleas from nesting animals, mites from rodents, kissing bugs from wood rats, and possibly soft ticks associated with larger animals. It's neat to have wildlife around your home, but not under your home. Animal tenants under porches and decks must be evicted, which sometimes requires the services of a trained technician from a pest control firm or a local government agency. Remember that, in many places, certain species of wildlife belong to the state so that special permission may be required to remove and relocate the animals.

There is a tendency to place shrubbery and trees directly adjacent to a house. Visually, the vegetation creates a border for the house and also creates a sense of distance and separation from the yard itself. Dense shrubbery adjacent to the house can cause two kinds of biting bug problems. First, the shrubs give rodents shelter, sometimes attracting them sufficiently close to the house that the animals naturally expand their activities indoors. Some kinds of vegetation, like English or German ivy, are particularly bad because they create a thick blanket of shelter. Almost any shrubbery encourages rodents when the plants are allowed to grow close to the ground and touch the wall of the house. It is a good idea to have one or two feet of open space between the wall and the bushes, providing air circulation and easy inspection.

A large tree adjacent to a house is sometimes the troublesome "cat" grown up from the cute sapling "kitten." The tree grows to a behemoth that would be expensive to remove and whose dangerously drooping branches cast a shadowy pall inside the threatened home. These are the same trees that fill the roof gutters with leaves, creating larval mosquito sites. It is not a common problem, but sometimes birds roost or nest on branches that touch the house. As long as there is physical contact, the mites can march from the birds to your house. Entering through any small crack, these kinds of mites can build up to large numbers inside the house. They'd rather stick with poultry, but the mites definitely find that humans taste sufficiently like chicken to be worthy of a bite. Such a problem can be solved rapidly by doing what is necessary for many other reasons: trim the tree away from the house.

No matter how different houses appear, they all have the primary purpose of providing shelter and living space for humans. The property on which the house rests can vary in many different ways. Open space might not exist at all on the property, or it might be many acres of land. The purpose of the space is also highly variable, depending on the family's activities, local customs, income level, etc. We can make two general statements about the biting arthropods that might come from the space around a house. First, some bugs coming from the property might enter the house and do their biting inside. In general, the greater the distance between the source and the house, the less chance of a problem. Second, some bugs will bite the occupants of the house outdoors as people garden, perform chores, or simply try to enjoy being in their yards. In this case, the severity of the problem will depend a great deal on the activities of the individuals.

Let's start with the biters that might fly indoors. Mosquitoes are the main problem, though many species of mosquitoes will not come inside. The yellow fever mosquito, Asian tiger mosquito, and some of their kin develop in many kinds of containers, but they also like to fly into houses to find human blood. Almost any container that holds water continuously for at least seven days might be the source: holes in trees, dishes of water for pets, old tires, unused boats, wheelbarrows, discarded cans, crinkled plastic sheeting, buckets, coconut shells, bamboo...well, you get the picture. The problem is easily solved in many cases by discarding or overturning the container. Containers that can't be drained (for example, tree holes, large cement containers) can be filled with sand, cement, or polyacrylamide gel (the stuff used to hold water for cut flowers).

The house mosquito will develop in containers, puddles, abandoned pools, or just about any long-standing accumulation of water with some organic matter in it. Containers are easy, but an unused swimming pool, the edge of a pond used by domestic animals, or even a low spot with persistent puddles may be much more difficult to eliminate. Any of a number of mosquito-eating fish can solve the problem sometimes. In the United States, many mosquito abatement districts provide fish for free. If you live in a tropical climate or if you want a solution for the summer only, many kinds of aquarium fish will do a good job (guppies, tetras, gouramis, etc.). Be forewarned, goldfish and koi do not eat mosquito larvae. You may also get hordes of house mosquitoes from larval sites that are not on your property. Two common sources are storm drainage systems that don't drain and industrial sites (like sewage treatment plants and factories that have wet processes) near homes. You may not be able to tackle these sources by yourself, but complaints to local authorities can cause effective changes in structures or procedures.

In some parts of the world, malaria mosquitoes can also develop right in your own backyard. There are species that like containers in South Asia and Latin America, species that like the margins of small streams in many parts of the world, and species that like the vegetated edges of ponds. Fish can be helpful, but they won't be nearly as effective when the sides of a pond are shallow or thickly vegetated. One of the worst transmitters of malaria in Africa likes to develop in thickly vegetated depressions that are often the result of normal household practices in that region. Tragically, a family plagued by malaria, possibly losing half the children to the disease, might be producing the mosquitoes on their own land. We probably do not know how many lives could be saved by adequate grading, cleaning ditches, and other measures that limit the extent of standing water in places where malaria is a problem.

Some of the malaria mosquitoes bite indoors, others outdoors, but almost all of the floodwater mosquitoes bite outside. As you will recall, these mosquitoes lay their eggs in the soil in places that are likely to flood sometime during the next year or two. Depending on your geographic location and the particular conditions of your property, you may be lucky enough to have a productive site right where you live. The larval habitat might be on your land or directly adjacent. From the individual mosquito's standpoint, the sites are all one thing—a place to lay eggs that is likely to remain flooded long enough for the larvae to complete development. From our standpoint, those sites come in a dizzying variety, representing many different natural and human-made situations. Among the kinds of places that support floodwater mosquitoes are the edge of a river that rises and floods a series of small pools, a low marshy area that floods after heavy rains or when snow melts in the spring, or the shallow, gently sloping edge of a pond that occasionally floods. Each of these situations presents tremendous challenges for a homeowner. Regrading, sometimes with underground drainage tiles, to eliminate standing water is the only practical means to permanently disrupt the larval site, though plowing the ground that contains the mosquito eggs eliminates some species. Unless a few shovels full of dirt will fill or drain an area, the job of regrading is likely to be a big one. What is more, the result might be an unpleasant change in an otherwise treasured natural wetland habitat. The right solution may be to watch for the larvae and then kill them with a registered, legal insecticide (see chapter 5, "WANTED: Dead or Dead").

Before leaving the subject of homegrown mosquitoes, we should touch on an issue that is likely to become increasingly important in countries with a lot of vehicular traffic and paved surfaces. Cars produce a certain quantity of toxic pollutants that wash around on roads. Hard surfaces prevent water from filtering through soil that might hold and detoxify these pollutants before they reach a stream or river. Most cities go to a lot of trouble to construct storm drainage systems that quickly remove water from a cityscape directly into a stream, making water pollution that much worse. This creates two environmental problems. First, the various pollutants go directly into important aquatic habitats; and, second, water flow and the resulting erosion of stream banks surge during rains. The engineering solution being applied to this problem involves a long series of recommendations aimed at reducing the amount of impervious surfaces in urban construction. Another type of solution is to direct the runoff into mini-wetlands that break down pollutants and slow the surge of water during rains. The mini-wetlands take many forms, from gentle swales in backyards and right-of-ways, to beautiful rain gardens of wetland plants, to large retention ponds. The design requirements (known in the trade as BMPs, for "best management practices") of these mini-wetlands include provisions to prevent mosquito development. As you can imagine, it is not so easy to design a naturalistic wetland that can't produce mosquitoes, but engineers have come up with some practical designs. The mosquito problem occurs when the mini-wetland is either poorly constructed or poorly maintained. Water that stands more than a week may create a floodwater mosquito habitat and water that stands for more than two weeks may create a permanent water mosquito (house mosquito, malaria mosquito, and others) problem. If the wetland is designed as a retention basin—forming what is often a handsome and productive habitat for plants, birds, and fish—it may require constant attention to be sure the balance leans away from mosquito production. For the professional environmental engineer, these sites are an interesting challenge, especially since they are necessarily located in urban areas where people are most abundant. As a homeowner, you can help yourself by being aware of the BMPs in your neighborhood and alerting local authorities when maintenance seems to be lacking.

Although unusual on a property smaller than a farm, black flies and biting midges can have a source near the house. A property that backs onto a stream or that is located on a slope above a mountain valley might have a problem with black flies originating close to home in rapidly flowing, clean streams. In this case, eliminating the source would involve a major engineering project that would probably be viewed as damaging. Better turn to chapter 5. Biting midges can originate from the damp soil at the margins of ponds, particularly ponds used by livestock. Digging steeper sides on the pond or fencing the area off from livestock are both difficult, but feasible, solutions. Another kind of biting midge develops in cracks in soil with high clay content. These black gnats can be impressive pests. Although the solution is somewhat experimental, you can probably limit the emergence of these beasties by plowing the cracks out of existence or by establishing an irrigated lawn in the area. Biting midges also plague some beach houses. In this case, eliminating the source is clearly impossible. In fact, there are no really good solutions to this problem, but you might get some relief from the suggestions in chapters 5–9.

Water is not the only source of biting pests in the backyard. Just like when we discussed roof gutters, we can get back into nagging mode by considering garbage, the maintenance of domestic animals, and general cleanliness. From the perspective of an industrialized, prosperous society, it is easy to feel critical about the accumulation of trash and food waste in a backyard. The fact is that much of the world does not enjoy organized trash pickup. Under those circumstances, each household does the best it can to store, compost, haul, bury, or burn its garbage and trash. It is a huge challenge to deal with it all in a way that avoids the production of pests, though there are some methods described below. The better job you do, the less you will suffer from homegrown pests.

Let's take a close look at a modern home with plenty of community services available: Rats like clutter and they like food. Generally, neater yards provide fewer hiding places that give rodents a chance to get started. Garbage should be stored in stout plastic or metal cans with tight-fitting lids, never in exposed plastic bags. If the city collects garbage two times per week, be sure to take advantage of both pickups. Rats may also eat fruit, either after it has fallen or when the fruit is still attached to the tree. Harvesting fruit as soon as it ripens can reduce this problem. Rats love pet food, chicken feed, and other animal chow. Provide your pet only as much food as it will eat in a few hours, and store animal feed in tightly sealed, metal containers.

So what about sanitation when there is no trash pickup? The same general rules apply: minimize structures that provide hiding places and make potential food sources unavailable. One strategy that works particularly well in tropical climates is to divide household garbage into four categories: recyclables, metal, flammable items, and food waste. Recycling may be as simple as taking the trouble to distribute containers to those who need them. Often, an enterprising person in the community picks up many different categories of recyclables or even pays a small price to get them. Just be sure that the paper, metal, bottles, or plastic do not accumulate in such a way that they provide mosquito larval sites or rodent harborage. Metal that can't be recycled can be separated from everything else and piled in a corner of the yard. By being careful to flatten cans or other items that might hold water, the potential for mosquito production is eliminated. In tropical climates, iron-based items disappear amazingly quickly: rust never sleeps. Flammable items are best burned in the open outside so that potentially toxic fumes from inks or plastics are not a problem. Of course, burning must be avoided if local regulations do not allow it. Food waste can really accumulate from a large household. In the absence of a garbage disposal and trash pickup, dealing with food waste requires some effort. If you have animals on your property, potentially all of the food waste will be eaten by the chickens, pigs, or cattle. Just be careful that the animals eat all of the food promptly so that rats do not have a chance to join in the meal. Compost heaps will consume a huge amount of food waste if they are properly maintained (and they will produce stable flies and other flies if they are not).

The world is awash with good advice on composting. Rotating barrels, multi-pile systems, worm farms, layered plastic devices—they are all immediately visible to anyone who picks up a gardening magazine. Obviously, if food scraps accumulate too quickly, the compost heap will become a rodent buffet. Turning the compost heap frequently keeps it cooking so that food becomes unacceptable to rodents and eventually breaks down into a perfect fertilizer for plants. A less obvious problem occurs when plant waste becomes too wet. At just the right moisture level, decomposing plant material by itself or mixed with livestock feces (horse apples are best) quickly produces stable flies. The source doesn't have to produce many stable flies in your backyard to create an intensely annoying problem. Even a few of these persistent biters will change your contemplative outdoor time into an ordeal of slapping at agile biting flies. Stable flies do not do well if the compost is turned frequently to accelerate the decomposition.

A few final words on sanitation, and then the nagging will stop. Tropical areas have two special and dangerous biting insect problems that are often associated with outdoor household practices. Some sand flies—the pesky little bugs that transmit the parasites causing Baghdad boil, espundia, kala azar, and other diseases—develop in rodent holes; at the bases of termite mounds, adobe walls, or other earthen structures; and near moistened soil in desert environments. Although entomologists have a poor understanding of these flies' exact requirements, it seems clear that some of the bugs develop well in our own backyards. Eliminating unused adobe or brick structures, stopping leaks from water containers or pipes, avoiding urination on the ground, and discouraging rodents may all go a long way toward reducing the number of sand flies near a house.

The other special biter is the chigoe flea. Even one of these charmers will ruin your day by burrowing directly into your skin like something straight out of a science fiction story. The conditions that lead to their abundance are even more mysterious than for sand flies, but chigoe fleas are clearly more numerous in some households than others. The association seems to be with pigs, sandy soils, and porous paved surfaces (like bricks). If you have a problem with chigoe fleas, you might start your solution by moving the pigs far from the house, replacing brick floors or paths with cement, and ameliorating sandy soil near the house with compost.

Ticks and chiggers—oh my! We end this chapter with that duo of arachnid pests, ticks and chiggers. Most people think of these pests as something you pick up while hiking, camping, or otherwise enjoying the great outdoors. The fact is that it is quite possible to grow your own right in your backyard. Most ticks require large mammals to complete their life cycle and the recent superabundance of deer in the United States has definitely benefited ticks. To make matters worse, the most important tick-borne pathogen in the United States, the Lyme disease bacterium, is transmitted by a tick that does very well in backyards that have occasional visits from deer. Excluding deer will only help if they are kept at a considerable distance, because rodents will carry the small larval ticks to your yard. The larvae drop from the rodents and mature into nymphs, the stage that most often transmits the bacteria to humans. You can reduce the risk of backyard tick infestation by keeping the ground dry. Cutting the grass short, trimming bushes to provide good air circulation under shrubbery, and excluding rodents will also help. Poison will work even better (see chapter 5). Chiggers also like shrubbery, especially berry bushes. Depending on your geographic location, either lizards or rodents support chigger populations. Thinning vegetation and discouraging their natural hosts are probably the only good solutions to this problem.

Stopping bugs at their sources is usually the best first step in stopping bites around the home. Despite your best efforts, however, you may find that the pests are still a significant problem. Of course, no matter what you do around your home you might also encounter biters in places you only visit as a tourist, worker, soldier, or passerby. Insecticides can be a helpful tool, killing those biting bugs that we can't keep from developing in the first place. Let's see how to use insecticides safely and effectively in the next chapter.

WANTED: DEAD OR DEAD

Let's assume that everything possible has been done around your home to make it the kind of place where bugs are unwelcome. If they ask to spend the night in your roof gutters, you tell them to find a house where the owner doesn't bother to clean them. If they say they won't be a bother, that they'll just use the gaps under your roof, you tell them that *your* roof is bug-proof. If they promise to leave you alone and only bite your rats, you say, "What rats?" Unfortunately, you may have cleaned, repaired, and excluded, but you may still have a problem with biting arthropods. The problem might come from places you simply cannot fix in your home. Or the problem may come from outside your home, usually on the wings of bugs that developed somewhere else. Outside of your own home, you have little chance of reducing the sources of annoying arthropods.

Insecticides can be useful tools for solving these problems, but many people have severe doubts. Insecticides are poisons that kill insects, mites, and ticks—often on contact. Anyone would be forgiven for thinking that, if the poison kills a bug, it might kill, or at least harm, a human. Some insecticides are indeed deadly and must be handled with extreme care. Entire industries are built around the challenge of distributing a toxic chemical on crops or other target sites without exposing the person doing the distribution or others in the vicinity. For these highly toxic substances, the strategy is to prevent virtually all exposure of the people to the chemical. Other insecticides are much less toxic to people than they are to insects. These insecticides are either broken down quickly by human metabolism, or the way they poison insects targets a physiological mechanism that is different in people. The vast majority of home-use insecticide products are in the latter category because governments and manufacturers know that John Q. Public is unlikely to follow the strict safety procedures necessary to handle the more toxic chemicals.

Public attitudes toward insecticides are sometimes very strong, even violent. It makes sense that a logical person would question the wisdom of spreading a poison within his or her living space. From a consumer standpoint, every insecticidal product has the same purpose: to kill insects. As a result, there is little sensitivity to the fact that the safety of each insecticide depends a great deal on its active ingredients. People sometimes have vague fears about accumulation of the poison over a period of time, about the chance that the chemical will cause cancer, and about subtle, unknown, and negative health effects. Particularly questionable is the often-held assumption that any exposure, no matter how small, will have negative cumulative effects. At the opposite end of the spectrum are people who do not seem to have enough respect for the hazards of insecticides. Sometimes, they feel that the product is designed to kill insects and therefore cannot harm people. More commonly, they are caught up in the immediate need to kill bugs and they fail to take precautions that would ordinarily be routine. Finally, the public just doesn't seem to get as excited about new insecticides as they do about new drugs. Many new drugs and vaccines get a lot of attention from the news media because people view these products as key improvements to their lives. New insecticides rarely get such attention, even though they might protect many more people than the latest drug.

The spray can of new insecticide on the store shelf represents the last stage in a long line of research and development. The process of inventing and developing the product has occupied the careers of smart people and used tens, or even hundreds, of millions of dollars before a single cent of profit is realized. Ironically, the hardest part of the development is also the cheapest: inventing the chemical in the first place. These days, a new active ingredient might be synthesized quickly by making molecular modifications to an existing pesticide, by modeling known pesticide molecules in a computer program, by replicating the chemistry of plants or microorganisms known to have defensive substances, or by the brute force screening of large libraries of chemicals. Physiological studies of arthropods, on the other hand, turn the pesticide discovery inside out by first building an understanding of how an insect lives, then designing molecules to stop those vital processes. It sounds like a betrayal of our insect friends, but a favored strategy for developing new insecticides is to learn how bugs live in order to kill them. This process actually produces chemicals that can be very specific for certain kinds of pests and safe for humans.

After discovery of the new chemical, the real work begins. The substance, which is often new to science, must be synthesized in sufficient quantities for practical tests of its usefulness. Toxicological tests early in development inform researchers on handling restrictions for safety's sake. Legal registration of the active ingredients requires a long battery of tests to evaluate their safety to the environment, wildlife, domestic animals, and people. These studies generate many thousands of pages of reports, some of which are dedicated to the documentation that the studies were conducted honestly and thoroughly. Regulatory agencies like the U.S. Environmental Protection Agency (EPA), World Health Organization (WHO), and European Union review those data carefully, often calling for additional studies. The process can require years to complete, just to be sure that the intended uses of the chemical are worth the risks it poses. The process is complicated and slow because it needs to be. A bad drug might kill the patient, but no matter how tragic the outcome for that individual, the drug itself is buried with the corpse. In sharp contrast, a bad pesticide can damage our environment for decades or even centuries.

Let's take a look at two insecticides that represent opposite ends of the spectrum, basically showing how bad they can get and how good they can be. The bad one is, of course, DDT. Despite the fact that DDT has not been used extensively in the United States since 1972 and has been used only in a few countries since 1980, the very name is often taken as a warning of the evils of synthetic chemicals in general and insecticides in particular. What's in a name? Apparently a lot, because

DDT still evokes fear and disgust among the public, though its full chemical name, dichlorodiphenyltrichloroethane, is hardly known.

The history of DDT is fascinating and woven tightly into many other trends throughout the nineteenth, twentieth, and twenty-first centuries. We start with Othmar Zeidler—not exactly a household name. Zeidler was a doctoral student in the laboratory of the more famous Adolf von Baeyer, who was one of the founders of the Kaiser Wilhelm Universität in 1872 following the Franco-Prussian War. Located in a part of France that was traded back and forth with Germany, the ancient university returned to France in 1918 under its original name, the University of Strasbourg. Von Baeyer was a luminary in the new science of organic chemistry, working his way through synthetic strategies that are still foundation stones of modern industry. Zeidler did his bit as a graduate student by adding some knowledge in the form of a thesis published as a scientific article in 1874, "Uber Verbindungen von Chloral mit Brom und Chlorbenzol," which we can roughly translate as "I found DDT!"

The history of the world would be very different had Zeidler noticed dead flies in his beakers and made the connection between DDT and its useful insecticidal qualities, but the discovery slept for many decades. Von Baeyer got a Nobel Prize in 1905 for his work on synthetic chemistry, and poor old Othmar Zeidler died in obscurity as an apothecary in Vienna. The fact is that biology, agriculture, and public health were not quite ready for DDT in 1874, even if Zeidler had noticed its amazing ability to kill insects. The big discoveries in medical entomology did not start until 1879 with Theobald Smith's demonstration of the first arthropod-borne pathogen (the one-host tick and Texas cattle fever). The intellectual climate of the late 1800s went into a fast boil as malaria (Ronald Ross, 1897), plague (Paul-Louis Simond, 1898), yellow fever (Walter Reed, 1900), lymphatic filariasis (George Law, 1900), and Rocky Mountain spotted fever (Howard T. Ricketts, 1906) were all shown to be caused by arthropod-borne pathogens.

There were some major successes against these diseases using primitive methods applied with obsessive organization. For example, William Gorgas conquered the threat of yellow fever in the part of Panama where the canal was constructed. Fred Soper eradicated the yellow fever mosquito from the entire country of Brazil by 1943. These and other successes showed the world that control, and even elimination, of vector-borne diseases was possible. Understanding the causes of some of the great killers of humanity and knowing that, in theory, they could be tamed must have frustrated public health professionals as major outbreaks of epidemic typhus (for example, over 500,000 cases in Serbia during World War I), plague (for example, 12 million deaths in India, 1896–1948), and malaria (for example, 100 million cases in India, 1933–1935) continued to occur. By the late 1930s, the world was more than ready for some solutions to these problems, and Paul Hermann Müller set out to find them.

Working for the J. B. Geigy Company in Basil, Switzerland, Müller started a project in 1935 to find a synthetic insecticide that would be cheap to manufacture anywhere, highly lethal to insects on contact, safe to apply, and persistent. He worked through a number of related chlorinated compounds and came up with DDT as the best one in 1939. The compound was tested against the Colorado potato beetle, an invasive pest in Europe, and then used as one of the tools to stop an incipient epidemic of louse-borne epidemic typhus in Naples in the winter of 1943–1944. General Dwight D. Eisenhower, in a secret radiogram, dramatically demanded DDT in North Africa to stop the spread of the same epidemic. Another epidemic of typhus was successfully blunted in Korea and Japan in 1945–1946 as suffering Koreans went home from work camps in Japan and suffering Japanese returned from 35 years of occupying Korea. These triumphs and many successes against malaria made Müller a hero. He got the Nobel Prize in December 1948, and Professor G. Fischer offered this statement of gratitude in the presentation speech: "Thanks to you, preventive medicine is now able to fight many diseases carried by insects in a way totally different from that employed heretofore."

The world was giddy with the possibility of reducing medical and agricultural pests to mere nuisances. Because it was cheap and easily handled by applicators, DDT was used lavishly for almost any insect problem. Among the statistics: 80

million pounds used in the United States in 1959, 4 billion pounds used worldwide between 1940 and 1973, and an estimate of a billion pounds in the environment in 1968. There were many successes: malaria eradicated from 37 countries with a total population of 728 million, the eradication of the yellow fever mosquito from 11 Latin American countries, and the ability to wage war without creating epidemics of typhus.

The DDT fiesta of the late 1940s and 1950s had its critics from the very beginning. Rachel Carson tried to express those doubts in 1945, but she could not find a publisher. In 1950, the U.S. Food and Drug Administration warned that the dangers of DDT were underestimated, and in 1957 it restricted the use of DDT near aquatic habitats. That year was something of a turning point in the fortunes of DDT. The U.S. Department of Agriculture decided to start phasing out use of the chemical and, just as significantly, the *New York Times* reported Nassau County's lack of success in restricting DDT even though citizens of the county had noticed the disappearance of their formerly abundant bird life. In 1958, a letter describing "a small world made lifeless" came to the attention of Rachel Carson, who by then was a well-respected science writer. She began a scholarly review of insecticide use and translated her findings into the clarion call of the environmental movement, *Silent Spring*, published in 1962. Carson was dead 18 months later—from breast cancer. The year after, Paul Hermann Müller died.

Silent Spring was quite a book, and it is well worth perusing even today. The tone is reasonable, it is easily read, and it is unapologetically negative about persistent insecticides like DDT. As Carson describes one environmental travesty after another, the modern entomologist first feels guilty about the history of the profession and then is relieved that we no longer seem to make mistakes on such a grand scale. The profession today largely practices Carson's recommendations to examine the subtleties of toxicology before registration, to use insecticides that do not harm the environment, and to maximize the use of biological control agents (insects or pathogens that control pests). *Silent Spring* may have been the beginning of the end for DDT, but the tide had already turned against the chemical. A series of government advisory bodies advocated ever-increasing restrictions on DDT's use throughout the 1960s. In 1970, the newly established Environmental Protection Agency took responsibility for pesticides from the U.S. Department of Agriculture, and in the same year the first countries (Cuba, Norway, Sweden, USSR) banned the use of DDT. DDT was either banned or severely limited in country after country through the 1970s. One of the last countries to eliminate the chemical was Mexico in 1990. DDT is still used for public health applications in some countries, but its current production is very limited. The story isn't over, however.

Outbreaks of arthropod-borne pathogens (24,000 cases of epidemic typhus in Burundi up to 1997; introduction of West Nile virus into North America in 1999; Rift Valley fever in Kenya and Tanzania in 2007) and particularly the continuous epidemic of malaria in Africa, which claims 2 million lives per year, have motivated a reexamination of the usefulness of DDT. Ironically, the international Stockholm Convention called for the complete elimination of DDT in 2004, whereas the World Health Organization continued to endorse indoor house spraying with DDT for malaria mosquito control (though with the condition that replacement chemicals should be found). Whether or not you agree that DDT has legitimate uses today, few would disagree that it was grossly abused in the 1950s, causing a significant global disaster.

Let's go from one three-letter abbreviation to another and take a look at DDT's opposite pesticide number: Bti, which stands for *Bacillus thuringiensis israelensis*. Plain old Bt was first registered in 1961 as a preparation for the control of caterpillars, but Bti was not discovered until the mid-1970s.

The story starts when Leonard Goldberg of the Letterman Army Institute of Research, San Francisco, went to Israel on a fellowship to work with Joel Margalit at Ben Gurion University of the Negev. Specifically searching for a pathogen that would kill mosquito larvae, they ran across a puddle full of dead house mosquitoes near Kibbutz Zeelim. They sent samples to Hélène de Barjac at the Pasteur Institute, who identified the abundant bacteria in the water as a new variety of Bt. The world badly needed a replacement for the harsh chemicals normally used to control mosquito larvae. The mosquito research community turned its attention to Bti in a big way, performing study after study that proved its efficacy. The Bti bacteria make spores containing protein crystals that consist of up to seven toxins. These toxins have the convenient property of exploding the gut cells of the larvae of mosquitoes, black flies, biting midges, sand flies, and only a few other related families of flies. With Bti, the mosquito control industry had a tool that was the environmental opposite of DDT: not persistent in the environment, completely nontoxic to all vertebrates and most beneficial insects, and unassociated with any kind of harmful chemical residues. First registered in the United States in 1983, it is now the industry standard for mosquito larvae control.

That is the tale of two insecticides. One, DDT, with a dramatic history tied to world events, spectacularly successful and spectacularly damaging. The other, Bti, came on the scene because two entomologists took a walk with their eyes open. Bti filled a particular need in the control of biting insects and looks like it will continue to serve in that role for decades to come without environmental damage. Pesticides are not all one thing; even Rachel Carson acknowledged that nonpersistent, safe products could have been fine alternatives to DDT.

When you have a biting insect problem, how do you choose a product that will be safe for you, your family, and the environment? Perhaps the first decision is whether or not you even need a pesticide. Previous chapters should help you to decide. In some cases, you may have been able to solve the problem completely by stopping the bugs at their source. The best uses of pesticides include application to clean up the bugs left after removing their sources or by using the pesticide itself to eliminate the source. It would be unrealistic to ignore the fact that people often reach for the pesticide first in order to get what they hope will be immediate relief from what they perceive to be an intolerable problem.

If you do need a pesticide, the best place to start the selection process is the label. A legitimate product should have a detailed set of instructions either on the container or on a small pamphlet. If the product does not have such a label, *do not use it*. In the United States, the European Union, Japan, Australia, and many other countries, the label represents a legal document that gives some assurance that a person using the product correctly will suffer no harm, hurt nobody, and not damage the environment.

The label should contain the following essential pieces of information: the active ingredients, the pests that can be controlled with the product, the places the product can be used, and the manner of application. There are many kinds of active ingredients, though relatively few of them are used by nonprofessionals. It is unreasonable to expect a person to read a chemical name and then make a decision about the appropriateness of the product. What is more, active ingredients constantly change so that a final list of those that are safe and effective is not possible. Box 5.1 lists many of the important ones as of this writing.

We can divide the most useful products into a few categories. First, there are the organophosphates and carbamates. These are less and less available in the United States, but they are still common elsewhere. Dursban™ (chlorpyrifos-ethyl), dichlorvos (2,2-dichlorovinyl dimethyl phosphate, or DDVP), diazinon, malathion, Baygon™ (this trademarked name has been used for a series of products that includes the carbamate insecticide propoxur, and it currently applies to transfluthrin), Sevin™ (carbaryl), and Actellic™ (pirimiphos-methyl) are a few of the names that may ring a bell for those who have been around for a while. They can be used safely, but they are all toxic to one extent or another, and it is even more important than usual to follow safe handling instructions.

Box 5.1 Indoor Household Insecticides

It's complicated to choose an insecticide, especially one for use inside your own house. Like cold medicines, shampoos, and analgesics, all the labels seem to claim something similar but somehow different. How to choose? Does it matter?

It certainly does matter, and you can help yourself out by thinking about the products in two different ways. First, there is the form they come in. Sprays, solutions, dusts—they all have their uses (Table 5.1.1).

The second thing to consider is the active ingredient. There are quite a few unpronounceable active ingredients, and the list changes over time. What's more, the applications change as manufacturers and researchers find out more about the active ingredients. Fortunately, the label tells you what you can and can't do with the insecticide. When you examine the many products on the store shelf, look for the applications listed in Table 5.1.2. If it doesn't appear on the label, you shouldn't do it.

Table 5.1.1 Forms of Insecticide Application for Home Use

Product Type	Description	Advantages	Disadvantages
ready-to-use solution	liquid for sprayer	safe, no mixing, flexible use	need sprayer
concentrate (liquid or powder)	dilute for sprayer	cheap, flexible use	must mix, need sprayer
pressurized spray	press to release spray	easy to use	limited volume
indoor residual spray	use of any type of spray registered for indoor walls	can solve mosquito, bed bug, other pests for weeks or months	big job to prepare house in order to avoid contamination
total release spray	all contents released at once	treats entire room	specialized uses
dust	insecticide on powder	reaches bugs deep in crevices	messy
solid vaporizer	volatile insecticide dissolved into solid	treats an entire area	constant exposure to occupant
chalk, paint, felt-tip	insecticide in drawing medium	easy to use, visible where treated	messy, less effective

Table 5.1.2 Common Insecticide Active Ingredients for Specific Locations in the Home

Indoor Application	Common Active Ingredients
around food	deltamethrin, d-*trans*-allethrin, esfenvalerate, phenothrin, prallethrin, pyrethrins
bedding, mattress	deltamethrin, d-*trans*-allethrin, methoprene, permethrin, phenothrin, prallethrin, pyrethrins
carpet	deltamethrin, d-*trans*-allethrin, esfenvalerate, methoprene, permethrin, phenothrin, prallethrin, pyrethrins, pyriproxyfen, resmethrin, S-bioallethrin, tetramethrin, tralomethrin
clothing	d-*trans*-allethrin, methoprene, permethrin, phenothrin, prallethrin, pyrethrins, resmethrin, tetramethrin
pet area	bifenthrin, cyfluthrin, cyhalothrin, deltamethrin, d-*trans*-allethrin, esfenvalerate, methoprene, permethrin, phenothrin, prallethrin, pyrethrins, pyriproxyfen, resmethrin, tetramethrin, tralomethrin
vehicle	cyfluthrin, cypermethrin, d-*trans*-allethrin, esfenvalerate, permethrin, phenothrin, prallethrin, pyrethrins, pyriproxyfen, S-bioallethrin, tetramethrin

The pyrethroids are a large family of compounds based on the structure of the natural pyrethrum extract of a kind of chrysanthemum (Figure 5.1; the unextracted plant product is called pyrethrum; the extracted toxins are called pyrethrins; and the synthetic analogs are called pyrethroids). Some are more toxic

Figure 5.1. Pyrethrum daisy (*Chrysanthemum cinerariifolium*; plant about 24″ or 60 cm high).

than others, but for all of them there is a large safety margin between likely exposure and toxic effects. They are the work-horses of current household insecticides. Some of the familiar names are allethrin, resmethrin, permethrin, deltamethrin, and tetramethrin. PBO (piperonyl butoxide) and MGK 264 (N-octyl bicycloheptene dicarboximide) synergize pyrethroids—making the pyrethroid relatively more toxic to the insect.

We mentioned Bti as particularly safe for humans and the environment. It is the only insecticidal bacterial toxin for biting bugs that is available to the public, but there are other biora-tional compounds that use the insect's own unique physiology against itself. These products have names like methoprene, Precor, hydroprene, and pyriproxifen (Nylar), and they all mimic an insect hormone that controls development (that is, growth regulators). They are extremely safe for humans and other vertebrates, but they can kill a wide variety of arthropods. New products based on plant extracts commonly used as flavorings (for example, thyme oil, d-limonene, eugenol) claim safety as their chief virtue.

The rodenticides are the final group of household pesticides we will discuss. These products are important tools in the fight against biting bugs because rodents are associated with many household pests. Most rodenticides are baits laced with powerful chemicals like warfarin, brodifacoum, or diphacinone. These compounds cause the animals to bleed internally. Though many of us do not like to admit it, rodents are quite similar to people. As a result, these toxicants are also very toxic to people, pets, bald eagles, snow leopards, and African elephants. The products available to the public (Box 5.2) have tiny percentages of the active ingredients in order to make them safe to handle. Obviously (at least, we hope it is obvious), it is very important to follow the safety precautions on the labels of rodenticides. Although the vast majority of rodent-killing products contain this kind of active ingredient, there are other products that vary from virtually nontoxic to humans (based on indigestible oils), to dangerously toxic (arsenic compounds), to downright deadly (phosphine from zinc phosphide). The new nontoxic baits may be worth a try, but it is hard to conceive of a situation where the nonprofessional should use the more dangerous compounds. Phosphine (generated when humidity reaches the active ingredient, a metal phosphide) is particularly dangerous because it kills with a penetrating gas. Poor storage conditions or use in the wrong place can have deadly results—for the homeowner.

We have used the word *safety* over and over while discussing pesticides. What do we mean? We mean using the product only when necessary in a manner that does not threaten the health of the person using it, the health of other people or domestic animals, or the health of the environment. It only makes sense that the first principle of safe use is to minimize contact, or exposure, between the chemical and the person, animal, or environment. For example, no one would think of spraying a pesticide, no matter how benign, directly in the eyes.

Box 5.2 Controlling Rats and Mice

After you've done what you can to make your home unwelcoming for rodents, you can get rid of the ones that are left with either traps or poisonous baits. Both tools have their places, but many people prefer traps.

Traps are cheaper than poisons because they are reusable. They are safer than poisons because they are not, well, poisonous. On the other hand, some rat-sized traps have their own dangers because they are powerful enough to break a finger or a child's hand. Everyone wants a better mousetrap, which has resulted in a huge variety of clever devices. Practically any of them can work, but some are easier to use, or hide the dead rodents' bodies, or have pretty colors, or LED readouts. Among the varieties available are the standard snap traps, glue boards, and battery-operated electrocution devices.

Probably the biggest disadvantage of traps is that their effective use requires considerable skill. The two elements of skillful use are the selection of baits and the placement of the traps. The bait must be acceptable to the particular population of rodents in your home. Think about what they have been eating and use those items as bait. In practice, you have to be ready to try a variety of baits if you are not having luck with your first choice. Rats can be very suspicious of new foods and of the traps themselves. One way to overcome this problem is to put out baits without traps for a while to get them used to that kind of food. In some situations, a particular food is very attractive and you may even catch two rats in the same trap. That attraction probably depends on the relative scarcity of that kind of food in the home as a whole. For example, a nice moist banana slice may be a big treat in a dry storage room, but have little appeal in an orchard with fallen fruit. Whatever bait you use, be sure to hook it onto the pressure plate of the trap (before you cock the trap) so that the rodent has to tug it (for the last time). Mice are good at gently nibbling the bait without triggering the trap, a problem that can be solved by using baits like dried fruit, which can be attached to the pressure plate securely.

The places to put traps are also tricky. In general, you want to put the traps somewhere quiet, under or behind something, and against a wall. It makes sense to put traps in areas where the rodents are active, as shown by their droppings or rub marks. Sometimes, you can get clever and put traps near rodent-highway choke points, as at the top of a beam or just inside an entrance hole. The classic position for a snap trap is perpendicular to a wall with the pressure plate toward the wall. Don't get discouraged if the trap doesn't catch anything the first night; the rats may be getting used to the sight of a new object before they dare to explore it.

Poison baits for public use are usually in a ready-to-use form with very low percentages of the active ingredients. There are few exceptions, but the poisons are just about as dangerous to humans and pets as they are to rodents. "Keep out of reach of children" and "Keep away from pets" are warnings that really must be heeded. Bait boxes are one way to separate the rat poison from people. These boxes force the rat or mouse to go around a baffle, which forms an opening too twisted and narrow for curious hands or muzzles, before reaching the bait. Some locations, like attics and crawl spaces, really are inaccessible. In places where there is no danger of accidental poisoning, paraffinized bait blocks or "throw packs" of bait might be safe.

Any poison bait presents two vexing problems. First, the rodents may die inside a wall or above a ceiling, making removal of their bodies impossible. The little carcasses can stink for several weeks before they finish decomposing. Sometimes, a dead rat will produce a single brood of blow flies—quite an inconvenience when they emerge in the house. The best advice is to bag and dispose of the dead rats you can find and then tolerate the stink from the others.

The other problem with poison baits is secondary poisoning. This happens when predatory mammals or birds eat the dying or dead rats, picking up a lethal dose of poison. It is ironic that poisoned rats can harm the population of predators that actually do the best job of regulating rodent populations.

There are dozens of active ingredients used to kill rodents. Table 5.2.1 lists some of the common ones available to the public. Not to nag, but it is *very important* to follow label directions.

Table 5.2.1 Active Ingredients in Household Rodent Poisons

Active Ingredient	Mode of Action	Advantage	Disadvantage
brodifacoum	2nd-generation anticoagulant	single feed	highly toxic
bromadiolone	2nd-generation anticoagulant	single feed	highly toxic
bromethalin	neurotoxin	single feed	highly toxic
cholecaliciferol	vitamin D overdose	safer for wildlife	multiple feed
chlorophacinone	1st-generation anticoagulant	safer	multiple feed
diphacinone	1st-generation anticoagulant	safer	multiple feed
warfarin	1st-generation anticoagulant	safer	multiple feed
zinc phosphide	phosphine gas stops respiration	very effective	highly toxic

Common sense can go a long way toward protecting you from exposure by avoiding fumes in the air (let the spray settle before entering the room), direct contact with skin (wear gloves and a hat, change clothes after spraying), and any chance of ingestion (do not store the product in a food container, always label containers, wash hands after using the insecticide). The label often gives valuable directions for safe use that you might not think of on your own. The label might say gloves are not required or that a respirator (a filtering device to prevent breathing fumes or spray droplets) is required. Pay particular attention to warnings about wildlife because some fish, birds, and mammals (not to mention your neighbor's bee hives) may be much more susceptible than people.

You should always give special attention to the protection of children. Kids can be very curious about pesticides. They like the activity, special equipment, bright labels, and even the noise associated with insecticide application. Products must be kept out of reach of even the most responsible child, who might on an impulse decide to drink the product or copy the adults by attempting to apply it. It doesn't hurt to let children know that insecticides are poisons to be avoided and to show them how to recognize poison symbols. If you have access by telephone to a poison control center, make sure you have the number handy. For the most part, careful attention to the label will assure safe use of the product.

Now that you probably feel thoroughly frightened and reluctant about pesticides, take a deep breath through a respirator, pick up this book with rubber gloves, and be comforted by the fact that, for the most part, registered household insecticides are among the safest pesticide products. Manufacturers and regulatory agencies know that the products will be used by nonprofessionals in areas that include children and pets. People can still get hurt with household insecticides, but regulation since the DDT days has made it a whole lot harder to get into trouble.

INSECTICIDES AS BARRIERS

One of the simplest strategies for using pesticides is to create a barrier of poison between you and the bugs. The idea is that a crawling insect will absorb a lethal dose as it walks through the chemical barrier. Of course, barriers of this kind only have a chance of being effective against crawling biters, but that includes a lot of the arthropods that give us problems around the house. The biting mites, like rat mites and fowl mites, travel from the nests or bodies of their favorite hosts to the humans and pets within a house. If you can figure out where they are coming from and also where they are biting, you can lay down a barrier of insecticide between the two locations in order to get some relief. A chemical barrier just a few inches wide is usually enough to kill these very small arthropods. If you have biting mites in your home coming from a bird's nest outside a window, try treating the windowsill and maybe a few inches of the interior wall surface around the window. If it's rat mites, you may get a sense that you are bitten in certain locations in the house. You can treat around that area or between that area and the probable location of rats. Barrier treatments for these mites are just a way of getting temporary relief while you pull out all the stops to get rid of the rodent infestation or the offending bird's nest.

People usually get tick bites away from home where they have little control over the population of these pests. We mentioned the brown dog tick that actually infests homes as though it were a cockroach, but barrier treatments would not make sense against this pest. The brown dog tick is likely to be abundant wherever the household dog goes. In Europe, the habits of the tick are a little different, and infestations may occur in all parts of the house, whether or not the dog goes there. Under these circumstances, it would be difficult to create a barrier without treating the entire building. Barrier treatments do make a lot of sense for one particular, very important tick situation. The Lyme disease tick depends on deer for its survival, and so as deer have become more abundant near suburban homes, this tick has come to infest the margins of yards where the vegetation is tall. The middle stage of this tick's development, the nymph, will wander from these areas even onto clipped lawns where people often spend time. No one knows how much Lyme disease is transmitted right in the backyard, but it is a disturbingly common possibility. For those lucky enough to live near open spaces, you can't treat the entire forest adjacent to your yard, but you can treat the marginal areas of your property.

Bed bugs are a lot like people: they commute each day from their homes to their place of work and they occasionally make a big move to a new city in search of better opportunity. Because of those habits, insecticidal barriers work particularly well against bed bugs. The bugs like to live in cracks and crevices either within the mattress or in furniture or woodwork near the bed. Each night, they must crawl across the room and across the mattress to feed on the sleeping human. Applying the right pesticide between where you sleep and where the bed bugs spend their days can get you a lot of protection easily, even if you are a one-night visitor to the bug-infested quarters. Make sure the bed has no direct contact with other furniture or the wall and then spray the edges of the mattress, the frame, and the legs of the bed. You won't have any direct contact with the insecticide as you sleep, and products registered for indoor use will not produce toxic fumes, but bugs will have to crawl across a deadly barrier on their way to work. This strategy will only work if the bugs are fully susceptible to the particular insecticide you use. Unfortunately, there is more and more resistance to the pyrethroid insecticides that are usually registered for indoor use and for use on mattresses. The world badly needs alternative insecticides for bed bug control.

Remember that, while you are sleeping blissfully protected by your chemical barrier, some of the bugs are busily finding seats in your luggage, ready for their all-expenses-paid trip to your next destination. That next destination might be your home, giving you a souvenir you could really live without. Although mothballs would probably keep the bugs out of your luggage, no one can really guarantee their effectiveness. Probably a better idea is to spray the outside of the luggage before you get settled in the infested room, concentrating especially on areas near openings into the bags. As for your clothing, maybe it's not such a great idea to store it in the drawers or closets of infested rooms: leave your clothes in your treated suitcases or tackle the job of creating chemical barriers between your clothes and the bugs. A nonchemical alternative is to carry plastic bags with you. If the bag is tightly sealed with a twisty or other closure, then bugs will not be able to get into luggage or clothing that hangs in a closet (unless the bugs have infested the clothes hangers—which sometimes happens).

Chemical barriers can also work against stinging bugs, including ants, centipedes, and scorpions. The same principles apply, usually spraying windowsills, door frames, and the areas near any opening leading into the room. Ants often get into all kinds of furniture as they forage for food, and some centipedes and scorpions seem to like to get into beds and sleeping bags. Effective barriers against larger-bodied arthropods like centipedes and scorpions must be feet rather than inches wide in order to assure that the bug gets a lethal dose.

Barrier treatments against spiders are usually more psychological than entomological, as spiders in temperate regions rarely crawl around in the places where humans rest. In the tropics, there are situations when a barrier treatment against spiders might be justified. If you are unlucky enough to have a tropical home that is attractive to spiders that construct hundreds of interconnected webs to form a huge communal net, a barrier treatment would probably keep them away. Certain kinds of tropical construction provide many hiding places for spiders on the outside walls. A barrier treatment in this situation may actually be better environmentally than eliminating the source of spiders. The barrier only kills those spiders that venture indoors, whereas total elimination of the spiders removes an important natural control on other pests.

Lice and kissing bugs are not the best targets of barrier treatments. When you think about it, lice go directly from person to person through direct contact (head lice and crab lice) or in clothing (body lice). Putting up chemical barriers will not affect their distribution, unless you consider clothing treatment as a barrier. Kissing bugs are large insects that would require a broad, fresh barrier to prevent entry or to protect sleeping occupants. These bugs are a real medical threat because they commonly transmit the agent of Chagas disease. The best bet is to get them out of the house entirely. Those species that come in from outdoors often fly, making a chemical barrier on a surface ineffective.

SPACE SPRAYS

Space sprays are aerosols of insecticide designed to kill flying insects. The clouds of finely divided droplets commonly come from pressurized cans, or "bug bombs," which were first produced in 1943 to combat malaria during World War II. These are amazing products because they kill insects in the air that people breathe. What is more, the chemicals do not break down immediately, but settle on every surface of the room. The insecticides have to be the safest kinds available and the directions for use carefully worded. Manufacturers have risen to the challenge of producing a wide variety of safe, effective products that are among the most popular of all household chemical supplies. The labels of many of these products allow you to spray them on surfaces as a barrier, as well as into the air to kill flying insects.

In general, the insecticide in space sprays kills flying insects for a short period while the droplets float slowly downward in the air. Outdoors, this is likely to provide only temporary relief as more biting flies or mosquitoes filter back into the area. Depending on the situation, temporary relief may be just fine—just don't expect the spray to solve any problems on a long-term basis. It is a different story indoors. The mosquitoes that find their way inside will have a hard time escaping the droplets of insecticide in an enclosed space. What is more, the droplets are likely to stay concentrated for a longer time indoors, where air movement is at a minimum. Many labels advise turning off air circulation systems, spraying the room by directing the aerosol toward the ceiling and then gradually down toward the floor, closing the door, and leaving the room for a period of time. Even a less thorough approach can be highly effective. If, for some reason, you find yourself in an area with malaria and you do not take drugs to prevent it, spraying your bedroom before retiring can be very helpful. Even if you sleep with a bed net, a little aerosol insecticide lowers the chance that an infected mosquito will give you that deadly bite on the elbow or other body part that flops against the netting.

Similarly, spraying during the day may reduce the chance of getting dengue from day-biting mosquitoes. Be sure that the place you spray is mentioned on the label of the product: some active ingredients cannot be used near food or in other sensitive areas.

We should at least mention the indoor and outdoor devices, known as "misters," that automatically produce a spray at periodic intervals. In a functional sense, these devices attempt to overcome the temporary nature of the effects of the usual space sprays. The first question about these devices is whether or not they are safe. In parts of the world where there is thorough regulation, government agencies have calculated total exposure and toxicity to conclude that the risk is acceptable. However, your exposure to insecticide will be far greater if it is being sprayed directly into your breathing space every 15 minutes. The next question is whether constant delivery of a space spray is ever really necessary. Common sense would suggest that the requirement for such frequent spraying indicates a problem better addressed by stopping the bugs at their sources. Some people may use these devices "just in case" there might be biting bugs, which seems like a waste of money and an unnecessary exposure to insecticide.

ELIMINATING SOURCES OF BUGS WITH INSECTICIDES

Chapter 4 was entirely dedicated to stopping bugs at their sources and advised that the elimination of the source was preferable to killing biting arthropods after they emerge. Sometimes, though, you can eliminate a source by applying a chemical. Of course, successful treatment requires some knowledge about the bug so that you know the what, where, and when of application. How do you know when you've done a good job? The answer is, when the biting bugs are gone.

Let's start with ticks. Inside the house, you might have soft ticks (the ones that feed quickly and, often, painfully) or brown dog ticks. The soft ticks are usually in the house because of a rodent infestation, though some kinds are associated with poultry. If rodents are the problem, then the thing to do is eliminate the rodents. We already talked about structural changes to discourage rodents, but traps and rodenticides can be useful ways to deal with them as well. We described barrier treatments for hard ticks that might infest your yard. The kinds of ticks that transmit Lyme disease have a two-year life cycle, with the happy result that a barrier treatment may disrupt the life cycle and achieve control long after the chemical application has disappeared.

The common household mites require animals as their real sources. If the problem is rodent mites, then we are back to the same issue of rodent control. If the problem is fur mites, you may be able to treat the pet dog, cat, or rabbit with the appropriate pesticide, eliminating the problem. There are no commonly accepted pesticides for the treatment of birds that support fowl mites. In that case, your only recourse is to discourage the birds from nesting directly on your house.

Chiggers live outdoors in association with lizards or rodents. Most people have chigger problems when they are away from home, while working or hiking in the wrong places. However, chiggers have been known to infest backyards, bringing nature a little too close to home. Wholesale extermination of wild rodents and lizards in the yard would get rid of the chiggers eventually, but in the meantime, you will get more bites. A better approach is to use an outdoor insecticide on the areas where you are getting the bites.

The cures for lice all aim at eradication (who would tolerate just a few lice?), and they all involve direct treatment of the human host. Drugstore shelves always contain products for the treatment of head lice and crab lice. Labels usually advise two treatments applied about 10 days apart. The first will kill the current crop of lice, and the second will kill the lice that hatched from the eggs (called nits) unaffected by the first treatment. The treatments are usually lotions rubbed into the hair, allowed to stay on overnight, and then rinsed off. The directions vary, so be sure to read the label for the product you purchased. Some people use fine-toothed combs to remove lice and nits, but chances are that you will have to supplement the combing with application of an insecticidal lotion. If one member of your family has lice, you will probably need to treat everybody. Otherwise, you may find that you cure one person only to reinfest the family from someone else. Some space sprays list the treatment of lice in bedding on their labels, but head and crab lice are seldom transmitted that way. Body lice are a completely different story. They infest clothing, which then needs to be treated with boiling water, high-heat drying, or a properly labeled insecticide.

Bed bugs are insects you'd like to completely eradicate. Barriers of insecticide, as discussed above, are a good way to avoid bed bug bites in temporary quarters, but you should have zero tolerance in your own home. When bed bugs are susceptible to insecticides, the problem is often solved by treating the places bed bugs like to hide. For the most part, sprays will not kill the eggs, making it necessary to retreat after about 10 days when the eggs hatch. One particular product, containing phenothrin and synergists in an alcohol base, is said to kill eggs on contact. Other pesticides registered in the United States for the treatment of bed bugs on mattresses and other places are pyrethrins, deltamethrin, cyfluthrin, permethrin, chlorfenapyr, diatomaceous earth, and limestone. The do-it-yourself approach is likely to work best when you catch the infestation early. Once the bugs are well established, they can infest every nook and cranny in your home, even behind electrical switch plates and in walls. At that point, you will probably need to pay a significant amount of money for a professional pest controller to make multiple visits until the bugs are all gone. Elimination can be complicated, especially in large, multifamily buildings where it is difficult to keep some residents from reinfesting the homes of others. Sometimes, all the residents are evacuated for several days while the entire building is fumigated with a toxic gas. Such fumigations are complicated because each resident must be careful to avoid taking bugs in their luggage and clothing out of the building. Not only would the bugs reinfest the building when they move back in, but some bugs might end up infesting the quarters where the residents stay temporarily.

Nonchemical control measures against bed bugs are also important. Targeted vacuuming with a powerful machine and a crevice tool can get many of the beg bugs out of furniture and crevices. Vacuuming lowers the overall number of bugs so that insecticides have a better chance of completely eliminating them. Many bugs survive inside the vacuum cleaner, however, making it necessary to empty the canister or place the bag into a bug-proof plastic bag before disposal. Individual items can be disinfested by exposing them to about four days of freezing temperatures or just a few hours of hot temperatures (140°F or 60°C). Hand-held steaming machines are sometimes used by professionals to kill all stages of bed bugs on furniture, draperies, and other places where the pests may hide. Heating is sometimes used to treat entire buildings or individual rooms.

Kissing bugs are another insect that should be completely eliminated from a home. These large bugs transmit a parasite in their feces that causes an incurable and chronic condition known as Chagas disease. The risk of getting this disease is simply too great to tolerate. Nonchemical control measures that eliminate hiding places are important, but indoor insecticidal treatment is more than justified. You may need a professional to apply residual insecticides in all the cracks and crevices where these dangerous insects hide.

The sources of mosquitoes are always water, but not always somewhere you can reach. It is usually preferable to get rid of the water that produces mosquitoes, as discussed in chapter 4. Sometimes, eliminating the water would be difficult (for example, a low spot at the bottom of a yard that would require the installation of drainage tiles) or undesirable for other reasons. Those other reasons might be that the water is used for a necessary household purpose or that it contributes to a desirable landscaping or habitat goal.

The pesticides available to treat mosquito larvae are limited because of the danger to the aquatic environment. Fortunately, three active ingredients are commonly available and very safe. In fact, they are so safe that they are legal to use in drinking water in many countries, though not in the United States. The first of these chemicals is temephos, often sold under the trade name Abate. The compound is amazingly toxic to mosquito larvae and amazingly nontoxic to people, though, as an organophosphate, it is chemically related to the most toxic pesticides known. The compound is not particularly selective for mosquito larvae, however, making it the wrong choice for a place you are trying to develop as an aquatic habitat. On the other hand, household containers, plant saucers, and small ornamental ponds would be good targets for temephos. A successful treatment may continue to kill larvae for 30 days or longer, especially if the chemical is prepackaged as 1% or 2% on sand granules.

Methoprene is another useful active ingredient usually sold in little carbon-based granules or in large bars. This compound works by fooling the insect's physiology into maintaining juvenile characteristics, eventually resulting in misshapen larvae, pupae from which adults don't emerge, or adults that die soon after emergence. Tiny amounts of this chemical are sufficient to kill the larvae. One application of granules may continue to control mosquitoes for a month or longer. Methoprene is completely nontoxic to vertebrates, but it may affect any arthropod. Methoprene is therefore safer than temephos, and it would be a smarter choice around fish. The granules or bars protect the active ingredient as long as the material stays dry. As a result, it is possible to put granules in a dry place that is likely to flood, confident that the product will kill mosquito larvae even months later when the site has water. The bars are designed to provide even longer residual action, particularly in places like cement basins where the chemical can distribute itself easily. It is important to remember that methoprene must be absorbed by the developing larva—it will have no effect on a pupa.

Bti, the third active ingredient for mosquito larva control, is available to the public in a number of forms. Granulated and briquette products have a long shelf life and claim an extended period of activity in the water, but the Bti toxin itself breaks down quickly in the larval habitat. One complication about using Bti is that it must be eaten by the larvae to be effective. Most species of mosquitoes quit feeding about a day before they pupate, making it possible to apply the material too late to kill the larvae. Despite this complication and its more delicate storage requirements (cool and dry), Bti is probably the most useful pesticide for mosquito larvae. It is completely nontoxic to most other arthropods and all vertebrates. You can confidently use it anywhere.

Three other chemicals are recommended for use against mosquito larvae in some parts of the world, but they are not yet commonly available to the public. Pyriproxyfen and novaluron are new active ingredients that, like methoprene, disrupt normal mosquito development. They can be used at incredibly low dilutions, extending the margin of safety for humans. However, both products need to be used carefully to avoid damage to aquatic habitats. The third active ingredient has been around for a long time and is usually sold under the trade name Agnique. It forms a surface film from long-chain alcohols (poly [oxy-1,2-ethanediyl], alpha-isooctadecyl-omega-hydroxy—say that three times fast!), which smothers mosquito larvae and pupae with no toxic action at all. This product is the only one that may be used in drinking water in the United States. The product is somewhat difficult to use as a liquid, but a new granular formulation offers hope that it will become a useful product for the public.

Fish can be a marvelous solution to a mosquito problem. The right kind of fish will eat all of the larvae they can find in a container or pond. The limitations are when the water won't support fish, when there is so much vegetation that larvae can hide, and when fish get scooped out of the container during normal use of the water. Not all kinds of little fish eat mosquito larvae, one of the common exceptions being goldfish. Guppies, betas, tetras, *Gambusia*, and many other aquarium fish will do a good job. Many mosquito abatement districts distribute fish for free. Take advantage!

Most kinds of biting flies develop in sites that are either very dispersed or inaccessible. Unfortunately, insecticides are not likely to eliminate the sources of horse flies, biting midges, sand flies, stable flies, snipe flies, and other biters. One exception may be black flies. If you have a stream on your property that harbors the larvae of these vexatious pests, a slug of Bti or temephos may eliminate them. Treatment of streams is a tricky business because what you put in the water at point A may be consumed by a person or animal at point B. Be sure to check the label and local regulations carefully before attempting any application of an insecticide to a stream.

Insecticides are a great help in flea control. Changing pet bedding, regular carpet vacuuming, and rodent control all help, but most pet owners sooner or later need chemical help to get rid of fleas. There are two strategies for controlling the fleas that commonly infest homes and pets. The first is to treat all sites where larvae develop with a form of methoprene (common trade name Precor). The methoprene lasts a long time indoors, making it impossible for the fleas to complete their life cycle in the house. These products are usually combined with a pyrethroid or pyrethrins to knock down the current crop of adult fleas, forming a complete and one-step treatment. This strategy is extremely effective as long as the source of the fleas is in the house and the application is thorough. Products in a total release can of spray are often successful, but sometimes fail because the spray only reaches surfaces that are exposed on the upper side to the downward drift of the material. Ready-to-use solutions or push-to-release spray cans can be directed under tables, chairs, or other places where the pet spends time feeding its fleas. Sprays also work well outdoors in warm climates where fleas can reproduce in sandy soil or other locations that protect the larvae. The chemicals may have to be reapplied more often because they do not last as long outside.

The other strategy is to treat the pet itself. There have been many home-based products for pets, including shampoos, collars, and "pour-ons" spotted on the animal's back. The most recent addition to the market is a pill given to the pet once a month. The insecticide is harmless to the pet, but makes its blood poisonous to fleas. In this way, any flea that the pet encounters is killed before it can establish a population in your home. Be careful to follow directions—it would be a shame to solve your flea problem by killing your pet.

Do you remember the other kinds of fleas? Human fleas do not depend on pets and they seem to like to spend more time on the clothing and skin of a person. Nonetheless, the larvae develop in much the same way as the pet-based fleas; therefore, the methoprene + pyrethrin kind of treatment should help. In addition, it would make sense to treat as you would for bed bugs because the human fleas will be around the mattress and associated furniture. A number of products are registered for use on mattresses against fleas and bed bugs. Sticktight fleas on poultry require approved veterinary treatment; these fleas are a nuisance for the poultry farmer but a real problem for the birds. Chigoe fleas also develop much as do other fleas, except that the female has the unpleasant habit of burrowing into skin to feed, then growing to the size of a small pea. Considering the pure misery of a chigoe flea "bite," little is known about their control. The larvae seem to be associated with sandy soils and pig raising, therefore an outdoor flea treatment in such an area might help if chigoes are a problem.

Stings from venomous arthropods are often avoided by avoiding the arthropod. Just like your mother told you, be careful where you put your hands! In some parts of the world, scorpions actually infest houses—probably in response to infestations of prey items like cockroaches. Avoidance is difficult if scorpions commonly use your living room as a hunting ground. An infestation of scorpions might warrant insecticidal treatment of the subfloor areas, closets, or attic, in addition to getting rid of the cockroaches. Bees and wasps, though highly beneficial in themselves, sometimes make nests too close to the home for comfort. After a few stings, anyone would be justified in changing their live-and-let-live attitude to one of how-do-I-get-rid-of-these-critters? In some communities, it is quite easy to call the beekeepers' association to remove a hive of honey bees. The bees are valuable insects and someone will actually want them. However, sometimes the hive is too big or it consists of difficult-to-handle Africanized bees (sometimes inaccurately called "killer bees"). In that case, it may not be possible to save the colony for commercial use and it will be necessary to call in a professional to destroy it. This is a job you should not do by yourself. Wasps, on the other hand, are easy to treat yourself. Just buy one of those products designed to spray them from a distance, wait until evening when the wasps are in their nest and inactive, and then blast them. Some species of ants sting, but any ant infestation in the home is usually unwelcome. The store shelves are full of ant poisons, but the first thing to try are the baits. Modern baits are highly attractive to ants and contain poisons that the ants carry back to the nest. Just be sure that you use the right bait for the right ant (some prefer protein, some prefer grease, and others prefer sugar) and also be sure to use enough bait stations. You can also carry the battle to the ant home front by placing baits directly on their mounds—a strategy that works particularly well with fire ants. There are products designed to drench ant mounds with insecticide, but these are less likely to be effective and they certainly use more active ingredient than do baits.

Removing sources and using insecticides can stop many bug bites, but sometimes it is necessary to take the next step of placing a physical barrier between you and the bugs.

GOOD FENCES MAKE GOOD NEIGHBORS

Pest problems often seem unfair. We ask ourselves why the ants feel it necessary to enter the kitchen, why the rats assume ownership of the basement, or why the mosquitoes move swiftly into the decorative pond. If only, we tell ourselves, if only the pests would stay where they belong.

A wide variety of barriers are available to keep insects on their side of the fence, and many of these barriers have become important parts of daily life in industrialized countries. Window screens, solid walls, screened vents, chemical repellents, and traps are all examples. Using barriers effectively can change an infested, miserable situation into one where pests are only an occasional problem. Some methods help even when you are away from your house and on the insects' home turf.

There are a number of reasons to use barriers instead of pesticides, or in addition to pesticides. First, physical barriers are often cheaper and more effective than pesticides, creating a nearly permanent wall that requires little maintenance compared to the frequent reapplication necessary to maintain a chemical barrier. Second, the right kind of barrier can be much more effective than a pesticide; consider a stout vent cover that absolutely excludes rats from a basement compared to the uncertainty of poison baits that may or may not kill rodents before they come inside. Finally, barriers are generally less lethal to insects in general, a great value if you are trying to preserve beneficial insects like bumblebees and dragonflies.

We can start with the most basic unit of structure, the wall. Modern homes in cooler climates have solid walls that completely exclude any insect on the outside from getting inside. Go to the tropics and you will find much more variety. A wall might be just as solid as elsewhere in the world, or it might be nothing more than a loosely woven screen that provides privacy and shade. Such loosely made walls are commonly constructed from branches, bamboo slats, woven palm fronds, or even large leaves. In such a structure, you might as well be sleeping outside as far as the biting insects are concerned. The addition of a layer of mud or plaster can seal such walls, but not everyone will take the trouble, and it has to be done carefully to exclude insects that might seek one of the occupants for a blood meal.

Biting insects can also come up through a floor. Modern homes usually have floors that are quite carefully sealed, but a tropical hut might have significant gaps between boards or strips of bamboo. Especially in an elevated house, those gaps will allow any blood-seeking insect to gain entry. Screened-in porches are not going to provide protection if the deck planks of the floor are not tight against each other. This can be achieved with tongue-and-groove construction or with a sealant.

A house consisting of nothing but solid walls and a roof would be insect-proof, but it would also be people-proof. Almost any house has multiple openings in the form of doors, windows, vents, and holes for pipes, dryer vents, electrical utilities, cable television, etc. These openings come in many shapes and serve many purposes. There is not a single answer to the question of how to seal them against the incursions of insects and rodents.

The screen door may seem like a humble part of the house, but it represents a lot of thought and design (Figure 6.1). The best ones have springs or closing devices that shut the door quickly behind the person entering the home. The door seats itself firmly against felt or rubber seals and may even have a strip of rubber at the top and bottom to exclude insects from the narrow gaps. Modern screen doors have features like decorative grilles to protect the more delicate screen, and a whole industry has arisen out of the need to replace or repair torn screens. Banging open and closed all day long, a screen door is actually a pretty amazing device. Almost any screen

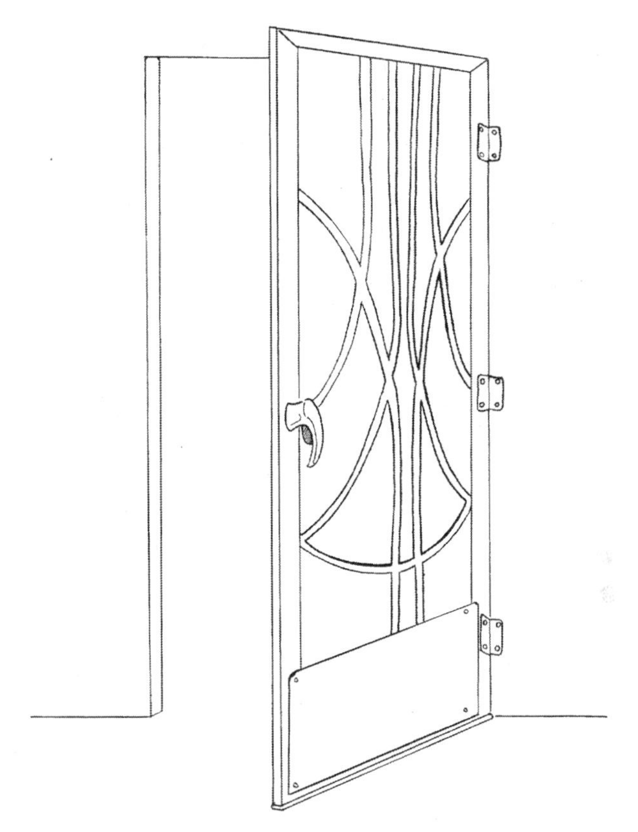

Figure 6.1. The humble screen door, a monument to technology for the prevention of insect bites. The kick plate, the seals on its edges, the fine-mesh screen, and the durability to withstand thousands of openings and closings. It's an amazing device.

door in good repair will provide some protection from flies, mosquitoes, and other flying insects. A screen door will also tend to deflect rodents that otherwise might wander in an open door, particularly if there is a solid kick plate at the bottom. No screen door is perfect, and some insects will literally follow a person into the house and later will find someone to bite. The house mosquito and the yellow fever mosquito are particularly adept at finding small gaps or imperfections in or around the screen door. The mesh size of the screen is also important. Sometimes, you see screens with a mesh size of 18 (each square mesh being 1/18″ on a side, or 18 holes per linear inch, equivalent to 7 holes per linear cm) or even the relatively gigantic 12 (5 per cm). Those mesh sizes are large enough for a small mosquito to enter. You want to be sure to use a mesh of at least 20 (8 per cm) and preferably 22 (9 per cm) or 24 (9.5 per cm). Biting midges and sand flies will pass through even these fine meshes.

Window screens are another standard feature of modern homes. Window screens vary in their sophistication from those that are integrated into $500 windows to nylon screens tacked onto window frames with wooden strips to seal the edges. A screened window will deflect wandering rodents, but if the beastie really wants to get inside, it'll chew right through. Good maintenance of window screens is essential because insects can enter homes through tears and holes. One common type of screen often has a gap between the aluminum frame and the window jamb, creating an entryway for mosquitoes even when the screen is in good repair. Mosquitoes, midges, and sand flies are going to tend to hit the screen that separates them from host odors, and then go down toward the gap that runs the entire width of the window. Small wonder that the gap between the screen's frame and the windowsill is often adorned with yellowing masking tape or a rolled-up towel intended by the occupant to stop blood-seeking bugs. Such seals will work, as will a layer of insecticide sprayed every week or so, but the best solution is to get a better window screen.

Other holes in the house are for utilities or for ventilation. Open gaps around wires, pipes, or dryer vents are the result of sloppy construction. The gaps should be sealed by masonry, epoxy compounds, or metal fittings sealed to the wall. Although a hole straight through the wall will allow some biting insects to enter, the bigger problem is from rodents. They can enlarge a small gap to get inside, and the pipe or wire becomes a superhighway. One handy way to stop them is to stuff the crack with steel wool, with or without cement. Rodents' ability to chew almost anything fails when confronted with the clinging fine wires of steel wool. Another common rodent route is a vent in the wall or just below the roof, designed to provide air circulation to attic, interfloor areas, or subfloors. The vents need to be covered with heavy 1/4″ (0.6 cm) screen (hardware cloth) or metal louvered grilles that exclude rodents.

In the tropics, it is not uncommon to leave a gap between the top of the wall and the roof. The gap allows hot air to escape from under the roof, particularly if the roof is not insulated by an attic space. Flying insects find easy access to the interior through these gaps. Although the gap can be sealed with screening, it is painstaking work to install the screen around the various supports and structures that interrupt the opening. Many households in the tropics do not bother with screens or screen doors at all. Partly a matter of cost, partly a matter of preference, some people think that a few mosquitoes and flies are less of a problem than the perceived reduction in air circulation caused by screens. When flies are numerous, some people complain that screens prevent the flies from escaping back out of the house!

Interior doors can make a surprisingly big difference to the distribution of flying insects in a house. Blood-sucking insects that get past screens or that enter as people pass through the front door find themselves in a situation where they must locate a host. Some mosquitoes are adept at seeking out the rooms where people sleep, traveling along hallways toward host odors. A normal interior door between the bedrooms and the rest of the house can reduce the problem dramatically—but only if the kids keep the door closed.

Bed nets are a great solution to a biting insect problem whether you are camping, staying in a hotel, or sleeping in your own bedroom. Bed nets cost very little, they can be used for years, and they provide nearly complete protection from all insects that can't pass through the mesh. People do not seem to use them as often as they should, possibly because the nets could be considered unsightly, they add one more chore to the bed-making process, and some people feel confined when they use a bed net. Although not a proven scientific fact, the reality is probably that people use bed nets when flying biting insects are enough of a problem to disturb their sleep. That's a shame, because consistent use would probably prevent more cases of diseases like malaria and West Nile fever.

Other than by color (Box 6.1), how do you choose and use a bed net? Mesh size is probably the first decision. If your concern is mosquitoes or kissing bugs, a mesh size similar to that of a window screen should work fine. The larger mesh

Box 6.1 What Color Is Best for Bed Nets?

A bed net is a big item in a bedroom, and it will be there every day. Each color has its advantages and disadvantages, some of which may seem trivial. However, bed nets are only effective if you use them, and you are more likely to use them if you like them as part of your bedroom decor.

White is often marketed to people who haven't previously used a bed net. The public associates white with cleanliness, Panama Canal–style sanitation, and the traditional unbleached muslin used for bed nets 100 years ago. White, like all light colors, probably presents a slightly less attractive visual target to mosquitoes. Light colors also make an easy background for spotting bugs that got inside the net before you did. The big disadvantage of light colors is that they show the dust and squashed mosquitoes, requiring frequent cleaning if they are going to look good.

Most nets for outdoor use seem to be some variation of dark green. In addition to the association of dark colors with ruggedness and the outdoors, they don't show the dirt and they offer somewhat better visibility from inside.

If you live where bed nets are a common part of daily life, you probably have a choice of a wide variety of designer colors. The lead author of this volume chose bright pink when he lived in South America.

size gives better visibility, allows more air circulation, and takes up less space. On the other hand, if you are trying to avoid the ravages of sand flies or biting midges, then you are going to want a much smaller mesh size. There are basically three general bed net designs: conical, box-like, and self-supporting (Figure 6.2). Conical designs are the commercial standard in the tropics because they use the least amount of material for the size of the bed. Another advantage is that they hang down from a single, central support over the bed; the top portion is held apart by a ring about one or two feet across. It's easy to arrange these nets neatly during the day by looping the bottom portion over the central support so that you have good access to sheets and blankets for bed making. The main disadvantage of the conical design is that the sides tend to drape inward, making more surface area that might come in contact with the sleeper. Wherever the net actually touches the person, there is no protection because insects will bite right through the mesh. Those experienced with conical bed

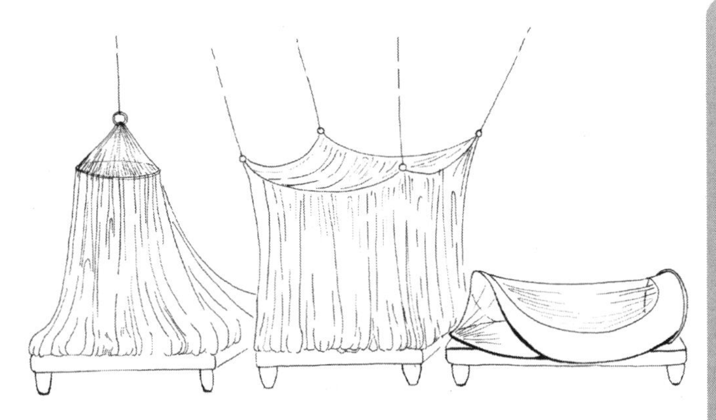

Figure 6.2. Three kinds of bed nets: conical, box-like, and self-supporting.

nets keep towels, extra pillows, or stuffed animals handy to brace against the sides of the net in order to prevent this problem. As with all bed nets, it is important to tuck the bottom of the net under the bed or sleeping bag. If you allow the sides to drape artfully toward the floor, insects will find their way through the gap between the bed and the net. In essence, if you don't tuck the bed net securely, you have created a giant mosquito trap with you as the bait.

Box-like mosquito nets have vertical walls and a flat top, reducing the tendency of the sides of the bed net to drape onto the sleeper. Nonetheless, if the bed is narrow, you will usually find a few mosquito bites on the elbows or other body parts that came in direct contact with the net. The box-like nets require support at each of the four upper corners, provided either by lines attached to something above the bed or by poles stuck into the frame of the bed. As you can imagine, box-like nets complicate bed-making a lot more than do conical nets.

Self-supporting nets have come on the market recently for outdoor use, though they would be handy for floor-sleeping guests or infants in a mosquito-infested home. They use the pop-up technology of tents, forming a tube-like enclosure with a floor. The entrance is usually through a zipper on the side or the end. Some have stout cloth near the bottom to prevent bites where the body might contact the sides.

Insecticide-treated bed nets have become a much studied and much discussed topic in public health since the mid-1980s.

They are being used as an elegant part of the solution for malaria, especially in sub-Saharan Africa. The idea is that an insect landing on a treated net is not going to bite, even if the net has not been tucked in properly, has holes, or comes into direct contact with the sleeping person. Nets can be treated with any of a number of pyrethroid insecticides, though only two of them (permethrin and deltamethrin) are currently registered in the United States for this purpose. The net can be dipped in a solution of the insecticide or sprayed, then allowed to dry like any laundry. The insecticide penetrates the fibers, binding closely to the net and lasting for months. A number of recent products, including one with permethrin in the United States, incorporate the insecticide into the fibers themselves, achieving effectiveness for more than a year and probably for the entire useful life of the bed net. Treated bed nets also help to prevent bed bug infestations.

Another kind of barrier can be important in the fight against bed bugs. Encasements are giant, bug-proof envelopes that surround the mattress and the box springs. Encasements can seal all the bed bugs inside an infested mattress, greatly simplifying the task of treating the rest of the bedroom. They can also be placed on new beds so that they cannot become infested. It is important to choose the right encasement specifically designed for bed bugs. Incredibly, bed bugs can escape between the teeth of some zippers or through the tiny gap at the end of the zipper pull.

Up until now, we have discussed the title of this chapter literally: physical fences between insects and people. In addition, a whole series of products attempt to create a harmless vapor barrier that keeps the bugs away. Though none are perfect, many are useful. The products are usually called spatial or area repellents, because they prevent biting within a volume of air occupied by people. Insects entering that volume of air either turn around to avoid the chemical, get killed or knocked down by the chemical, or lose interest in blood feeding. The chemicals themselves have to be very safe because, by the nature of their use, people will breathe them in.

In order to be effective, an area repellent product must be dispersed in the air as a vapor. Some chemicals are sufficiently volatile or evaporate rapidly enough to be effective at room temperature. For example, people have used mothballs scattered around outdoors to get some relief from mosquitoes. That is not such a great idea, because no one knows the effectiveness or the environmental safety of using mothballs that way. Several compounds were used in the 1940s as area repellents sprayed on the ground, but none of them are available now. One current product that works pretty well contains a mixture of chemicals absorbed into vermiculite. You can spread the product on the ground where you will be or in a 15-foot-wide band around the area. It takes about an hour for the product to be fully effective and then it lasts at least several hours after that. The product is well suited for a picnic or barbecue, but it is bulky, requires some care to distribute it evenly, and can only be used outdoors.

An entirely new product works by releasing vapors of a volatile pyrethroid, metofluthrin. There are several methods of dispersing the metofluthrin, the simplest being paper or plastic devices designed to maximize surface area for release of the chemical. They are simply hung up where flying insects are a problem.

No discussion of passive vaporization devices would be complete without mentioning DDVP strips. These have been around for about 30 years and have become a commonly recognized household item. They consist of a thick yellow plastic strip soaked in the organophosphate insecticide DDVP. Ironically, they are the only vaporization device registered for indoor use in the United States against mosquitoes, in spite of the fact that they contain an active ingredient much more toxic than any of the household pyrethroids.

A number of area repellent systems overcome the problem of dispersing a relatively nonvolatile chemical by blowing air over it or by heating it. The products that take advantage of these strategies are evolving rapidly. It is hard to summarize or evaluate the products here as they appear and disappear on the market rapidly. The principles behind them are sound, but some definitely work better than others. So far, the ones that blow air to disperse a chemical are generally less effective than those that heat an active ingredient. Part of any product's performance is due to the selection of the active ingredient (pyrethroids work best), but the actual amount of chemical released into the air is also a prime consideration. Heating the chemical is likely to put more of it in the vapor phase.

The oldest method of creating a barrier against biting insects is the burning of plant materials. Twenty-five hundred years ago, Herodotus described the ancient Egyptian practice of burning castor oil to discourage mosquitoes. A Roman compilation of daily wisdom, the *Geoponika*, recommended burning just about anything to stop mosquito bites. Smoke of any kind discourages biting, possibly by reducing the amount of water in the air and generally disrupting the chemical trail that leads to a host. Some plants probably create a much more effective repellent barrier when they are burned, as particular chemicals are vaporized. Local practices display a wide variety of species selected for this purpose (Table 6.1), but a few have been tested scientifically. The lemon eucalyptus, *Corymbia citriodora*, contains a number of chemicals known to repel mosquitoes, including one that is particularly effective. Burning neem (*Azadirachta indica*) leaves, lemongrass (*Cymbopogon nardus*), or Siam weed (*Eupatorium odoratum*) also repels mosquitoes. The low cost and availability of local plants makes burning an attractive option for people who do not have the means to use something more effective. Where an open fire is a routine part of daily life, burning plants known to have repellent qualities makes sense. The practice has two dangers, however. First, none of these methods is 100% effective; therefore, the assumption that there is complete protection from disease is very much mistaken. Second, chronic exposure to smoke of all kinds leads to respiratory ailments varying from irritation to asthma and from coughing to cancer.

By far, the most common area repellent is the mosquito coil. These clever devices consist of an insecticide-treated composite shaped in a spiral that is lit at the end. It burns without flame for up to eight hours. Originally developed in Japan as a modified stick of incense by Eiichiro Ueyama in 1890, his wife convinced him to develop a longer-lasting spiral in 1895. He used natural pyrethrum powder as an active ingredient. Pyrethrum is still used in some mosquito coils, as is a whole list of synthetic pyrethroids (Table 6.2). Coils have been immensely popular for 100 years and remain the most common form

Table 6.1 Examples of Plants Used by Local People to Repel Biting Insects

Material Burned	Where Used
black cumin (*Nigella sativa*)	ancient Rome
bushmint (*Hyptis*)	Brazil
castor-oil plant (*Ricinus communis*)	ancient Egypt
churai (*Daniellia oliveri*)	West Africa
coconut husks and papaya leaves	Solomon Islands
creeping thyme (*Thymus serpyllum*)	Russia
devil's dung (*Ferula asafoetida*)	India
eucalyptus (*Corymbia*)	Africa
fringed sagewort (*Artemisia frigida*)	Blackfoot tribe of North America
galbanum (*Ferula gummosa*)	ancient Rome
lemon bush (*Lippia javanica*)	Africa
marigold (*Tagetes*)	Zimbabwe
mint (*Mentha*)	Brazil
motacú palm (*Scheelea princes*)	Bolivia
neem (*Azadirachta indica*)	India, Africa, South America
vaca or sweetflag (*Acorus calamus*)	India
waste materials and smoke in general	worldwide
wet logs	Colombia
wild spikenard (*Hyptis suaveolens*)	Africa
wormwood (*Artemisia vulgaris*)	China, Central Asia, Bolivia, India, North America
yarrow (*Achillea millefolium*)	Colville tribe of North America

Box 6.2 All about Neem

Neem is a great example of a botanical source of many useful products. The neem tree, related to mahogany and chinaberry, is originally from India. Now you can find it in tropical locations all over the world. It grows to decent shade-tree size (up to 35 m), providing both food for livestock and lumber. Recorded uses of the plant go back at least 2,500 years. A seventeenth-century Sanskrit book, the *Yoga Ratnakara*, already called it nimba or neem. Other names for the tree are rempu, limba, margosa, sarva roga, nivarini ("curer of all diseases"), Indian lilac, balnimb, and its scientific name, *Azadirachta indica*.

In India, and increasingly elsewhere, personal care products often contain extracts of neem seeds. Toothpaste, hair creams, hand lotions, mouthwashes, and similar products include the extract primarily for its antiseptic properties. Simple extracts of the leaves have been used extensively as a cheap insecticide, mainly for agricultural use. These preparations are simple to make, but they produce complicated mixtures of at least 35 different chemicals. The extracts have the net effect of blocking essential insect physiological processes, stopping normal development and reproduction. The credit for most of these biological effects goes to one of the abundant compounds in the extracts, azadirachtin. Azadirachtin was registered as an insecticide for crops and ornamental plants by the U.S. Environmental Protection Agency in 1985. Like so many other botanical extracts, processing can produce a series of chemicals soluble in water and another series that forms an insoluble oil. The oil, which does not contain azadirachtin, has biological effects of its own, including the ability to kill fungus.

Neem products often target biting pests, and they include lotions that kill lice. Neem repellents for application to the skin are certainly available, but quantitative tests have generally shown them to be less effective than the best synthetic and botanical products. Neem frequently serves as a spatial repellent. The oldest applications simply burn or heat the leaves of the tree. More recently, the oil has been used as an active ingredient in heated mats or mosquito coils. One study showed that the addition of neem oil to the fuel of kerosene lamps significantly reduced the number of biting mosquitoes indoors. These methods show some promise, but they do not approach the effectiveness of natural pyrethrins or synthetic pyrethroids.

Neem is also a great example of the difficulty of safety evaluation when dealing with complicated botanical extracts. Environmentally, neem extracts kill aquatic creatures in the parts-per-million ranges of concentration. Azadirachtin itself has a low toxicity to people, but neem oil can be dangerous. The mixture of chemicals interferes with cellular respiration, resulting in a wide range of signs and symptoms, sometimes ending in death. It is estimated that a person weighing about 120 pounds (55 kg) could safely consume less than a drop (0.11–688 micrograms) of neem oil per day, in contrast to over a thousand times as much aqueous neem (825 mg). The tropical world recognizes the usefulness of neem by planting the tree everywhere it will grow, but neem must be used with the same care as any chemical product.

Figure 6.2.1. Neem tree and multileaflet leaf (*Azadirachta indica*; tree up to 130′ or 40 m high, leaf up to 16″ or 40 cm long).

Table 6.2 Effective Active Ingredients in Mosquito Coils

Active Ingredient	Comments
pyrethrum	natural, powdered material from a kind of chrysanthemum; moderate to excellent performance in coils
pyrethrins	extract of insecticidal chemicals in pyrethrum; works well in coils
allethrin (sometimes d-*trans*-allethrin)	the first synthetic pyrethroid; common active ingredient in modern mosquito coils; works well in coils
esbiothrin	a form of allethrin
prallethrin	another pyrethroid; known for its relatively high volatility
transfluthrin	another pyrethroid; known for its relatively high volatility
permethrin	another pyrethroid; somewhat persistent; toxic to cats; less suitable for coils than other pyrethroids listed
DDT	formerly a common ingredient in coils; *do not use*
lindane	formerly common in many kinds of coils and heaters; *do not use*
dibutyl hydroxyl toluene (BHT)	additive sometimes used to protect pyrethroid from oxidation as coil burns
piperonyl butoxide (PBO)	additive sometimes used to increase effectiveness of a pyrethroid
N-(2-ethylhexyl)-bicyclo-(2,2,1)hept-5-ene-2,3-dicarboximide (MGK 264)	additive sometimes used to increase effectiveness of a pyrethroid

of personal protection from mosquitoes and other biting flying insects. Some 29 billion coils are sold each year, and India, for example, spends 43% of its household insecticide budget on them.

Why are mosquito coils so popular? First of all, they work. A mosquito coil smoking away under a patio table in the evening can make the difference between swatting between drinks and sipping your piña colada. In quantitative tests, they provide something like 80% protection. Second, mosquito coils are cheap—and their cost is scaled to local economies. Their most expensive component, the pyrethroid insecticide, is used in tiny quantities, accounting for no more than a few tenths of a percent of the coil. Costs vary a great deal, but they range from less than 10 cents per hour to no more than 50 cents per hour. What is more, the coils can be bought individually and need no equipment for their use other than a match. Finally, mosquito coils are easy to use and fit into normal household practices of lighting candles or incense.

The main disadvantage of mosquito coils is that they produce smoke. Used indoors, the coil generates an unpleasant amount of smoke that is probably unhealthy. Even outdoors, one would probably like to associate soft tropical evenings with the smell of jasmine instead of the smell of burning sawdust and pyrethrum. And, like all area repellent systems, the effectiveness of mosquito coils is heavily dependent on concentration. Wind, fans, and distance all make mosquito coils less effective.

Devices that heat a chemical to produce repellent vapor come in many different forms, which vary widely in effectiveness. The least useful ones use active ingredients like citronella, sandalwood oil, or other botanical extracts. Such nice-smelling products are commonly formulated into candles—easy to use, cheap, nontoxic, pleasant, and…ineffective. This is not to say that effective candles with botanical active ingredients will never be invented, nor even that somewhere out there is not one that works pretty well. The recent introduction of candles containing a particular form of linalool may provide as much protection as most mosquito coils. The problem with most of the candles is that documentation is sparse and the documentation that exists shows that many candle products don't work.

The products that actually prevent most mosquitoes from biting contain pyrethroids. Among the common active ingredients are the more volatile compounds like allethrin, prallethrin, and transfluthrin. One of the clever ways of dispersing these compounds does not involve a commercial product at all. Studies have been performed showing that oil lamps used for light in rural African homes can be outfitted with small pans that heat a solution of insecticide (transfluthrin) to prevent most bites within a hut. Even adding insecticide (esbiothrin or neem oil) to the kerosene fuel can prevent many bites. These are not registered uses of these products, but the idea is certainly worth exploring in locations with simple living conditions where oil lamps are used routinely for light.

There is a whole series of products that use a paper mat soaked in insecticide as the source of area repellent. Insecticide vaporizes slowly from the mat, creating a cloud of chemical vapor typically about 15 feet across. Flying insects either avoid the vapor or die in it, dramatically reducing the number of bites suffered by a person in the middle. Obviously, wind tends to disperse the protective bubble, though some products work pretty well if placed upwind in a light breeze.

One of the first pyrethroid insecticides, allethrin, is a common active ingredient in paper mats. Allethrin is a very safe compound, but if you are in a country where the registration of pesticides is not required, you might encounter some active ingredients that are anything but safe. Lindane, for example, used to be a standard vaporized insecticide, but is now known to be quite a dangerous chemical. Avoid products that are not registered or that contain ingredients you don't recognize. Remember, you will be breathing fumes from an area repellent for a long period of time and you want to be especially sure that they are not toxic. Some labels restrict the use of mats to the outdoors, intending to provide a margin of safety for exposure to the chemical. All of the devices deliver a gentle heat to the paper mat, either with a candle, a butane-powered catalytic converter, or an electrical element. Prices vary dramatically, but the more expensive systems are about $1 an hour to operate, making cost a real factor in deciding whether or not to use them.

Wouldn't it be nice if you could grow a potted plant that kept the mosquitoes out of your house? Indeed it would, but the commonly sold mosquito or citrosa plant (*Pelargonium citrosa*, a kind of geranium) is virtually ineffective. Some outdoor studies showed that people sitting next to the plant got more bites than people sitting in the open. There is hope, however. Work on some African plants related to basil (*Ocimum suave*, or Kenyan tree basil, and *Ocimum kilimandscharicum*, or African blue basil) showed that 10 plants in a room could reduce mosquito bites indoors by up to 50%. Potted saplings of lemon eucalyptus (*Corymbia citriodora*), the botanical source of a very effective insect repellent, were similarly effective. Although growing indoor plants to prevent mosquito bites sounds like an elegant solution, 50% protection is not impressive.

Homes in subtropical and tropical areas are often surrounded by vegetation that provides refuges for biting insects. In Queensland, Australia, and Florida, United States, scientists tested the application of insecticides (bifenthrin and permethrin) onto vegetation as an additional barrier. The studies showed that fewer insects were found in treated homes compared with similar untreated homes, and the treatment remained effective for six weeks. In Australia, where biting midges (no-see-ums) severely affect the favored outdoor lifestyle in some locations, householders have their fences and vegetation sprayed with insecticide (bifenthrin) by licensed pest control operators to reduce the effects of biting insects. We discussed the use of barrier sprays in chapter 5, but mainly with the objective of eliminating part of the population. The same application can sometimes be used to inhibit the movement of the biting insects, therefore making a barrier much like screens or area repellents.

If good fences make good neighbors, what about mosquito traps? Any mosquito killed in a trap is a mosquito that does not find its target. A perfect trap would sit next to you like a good companion and attract all the insects before they could bite. Well, true friends are hard to find and so are those traps that act like good companions. There are problems inherent in designing a trap that can compete with a human host for the attention of a blood-sucking insect. The insects' quest for blood is the sharp end of an evolutionary spear that extends back millions of years. They are good at what they do, and they use multilayered systems of visual, physical, and chemical cues to get to the point of the matter. A trap that uses only light or only carbon dioxide has little chance of getting the attention of a mosquito when a big, sweaty, breathing human full of blood is nearby. Blacklight "bug zappers" simply do not work well, sometimes attracting more mosquitoes into a backyard and increasing the problem. More sophisticated traps that combine chemical and physical cues (warmth, water vapor, octenol, even heartbeat sounds) catch some biting insects more efficiently than others, but at least they catch them. In the right place against the right kind of insect, the right trap could make a difference. Just don't expect it to be that good companion wherever you go.

Traps are a more promising solution for stable flies, horse flies, and tsetse flies. There are numerous practical examples of the control of these insects by using traps. These larger

Table 6.3 Summary of Barriers Used to Stop Biting Arthropods

Method	Effectiveness	Pests	Comments
walls	absolute	all	must be solid without holes for wires, pipes, etc.
doors	excellent	all	some pests enter around sides and when door is opened
screen doors	excellent	all	structure, design, state of repair, and use influence effectiveness
window screens	excellent	all	gaps around edges and poor state of repair can reduce effectiveness
bed nets	excellent	all	important to tuck bottom of net under mattress, keep in good repair, and avoid touching net during sleep
passively vaporized chemicals	useless–effective	flying pests	outdoors only; many ineffective products on the market; work best when no wind; naphthalenes and metofluthrin are effective
chemicals distributed by air current	useless–moderate	flying pests	many ineffective products; metofluthrin works; could be better products in the future
smoke	moderate	flying pests	plain smoke helps; particular plants more effective; general respiratory hazard
candles	poor	flying pests	easy to use; ineffective
lamps	moderate–excellent	flying pests	home remedy in Africa; chemical added to fuel or in pan above flame; general respiratory hazard
mosquito coils	moderate–excellent	flying pests	outdoors only; work best when no wind; pyrethroids best
heated mats	moderate–excellent	flying pests	outdoors only; work best when no wind; convenient; sometimes expensive
live plants	useless–moderate	flying pests	commonly sold "mosquito plant" does not work; some examples in Africa of effective use
barrier sprays, misters	moderate–excellent	all	variety of chemicals work; expensive and labor intensive; kills many nontarget insects
traps	useless–moderate	small flies	generally do not meet expectations; could be helpful if enough good traps used; blacklight electrocuting devices useless outdoors
fly traps	moderate–excellent	stable, horse, tsetse flies	some traps commercially available are effective if used correctly

biting flies literally look for their prey from a distance, focusing on large dark objects that move. Apparently, they can also be sensitive to objects that show contrasts between dark and light areas, particularly if the sizes of the light areas are similar to the size of gaps that might appear between the legs of a large animal. The world seen through the eyes of a fly must be a very different place because objects like a swinging black ball a foot or two in diameter, a black cone, or a square of blue cloth with a rectangular hole cut in the bottom are unbelievably attractive to some of these flies. Traps that use these designs will collect flies by the bucketful, eventually depleting the entire population. Commercial trap designs for stable flies and horse flies in the United States make a significant difference even when only a few of them are used in a pasture. The main problems associated with the traps are the labor associated with emptying the flies and keeping the structure in repair, but this seems like a small price to pay for an effective, nonchemical solution to what would otherwise be a vexing problem.

Barriers against biting insects have absorbed a lot of human creative energy (Table 6.3). Barriers like screens and mosquito coils are used routinely by billions of people, but few of those people would claim that they have completely escaped biting insects. Despite literally centuries of technological development, we are still waiting for the perfect trap, the perfect area repellent, or even the perfect screen door. In the next chapter, we'll get a little more intimate and discuss how fashion can help to prevent a few bites.

SUIT UP!

Clothes might be said to separate humans from animals, but more important, clothing separates our precious skin from bugs. The right clothing can reduce the number of bites you get, even if there are many mosquitoes, ticks, black flies, or other pests around. The way you wear your clothing can also make a big difference. If you expect a really bad problem, there are special clothes designed specifically to stop bites.

Before trying on a few anti-bug fashions, let's take another look at the biting habits of our favorite pests. From the standpoint of preventing bites with clothing, there are three kinds of attacks. First, there is the insect infantry. They stay on a surface, crawl through the weeds (or perhaps the hair on your skin), and bite when they find a target of opportunity. Ticks, chiggers, mites, lice, and bed bugs all follow these infantry tactics. The advantage for them is that they can easily get access to your entire body once they find a place to enter under clothing. The disadvantage for them is that none of them can actually bite through clothing.

The second type of attack is like an air assault: bring the troops by air, then hit the target from the ground. Kissing bugs, black flies, biting midges, and fleas are their own helicopters, but they often crawl around a bit before they bite. You might easily find their bites just under the edge of clothing or hair. They flew (or, for fleas, jumped) on you, then walked around looking for just the right place. Kissing bugs seem to be careful to avoid disturbing their blood meal source by landing nearby and delicately biting with only their sucking tube in contact with the skin.

The third type of attack is more akin to something the air force would do. Mosquitoes, horse flies, snipe flies, stable flies, and tsetse flies tend to land right on the spot where they will bite. They might adjust fire slightly (mosquitoes often leave a line of welts as they search for the perfect spot), but they almost never wander far without taking flight again. With the exceptions of horse flies and snipe flies, they are pretty good at manipulating their sucking tubes right through cloth.

The infantry insects are perhaps the hardest to stop with clothing. They can crawl from the outside surface of clothing to an opening where they gain access to skin. Lice are a special case, in that they never leave the hair (head lice, crab lice) or clothing (body lice). Far from being protective, clothing and head gear are common sources of these pests when people wear each other's things. A clothes dryer on the hot setting, a vat of boiling water, or other sources of heat will kill all lice and their eggs, but there are other ways to get them out as well. Body lice, in particular, have been the target of various clothing treatments for a long time. As early as World War I, people experimented with washing underwear in a mixture of cresol and soap. World War II saw the introduction of the pyrethrin-based MYL powder and then DDT. These treatments revolutionized warfare by eliminating the deadly scourge of louse-borne typhus. Changes in textiles have also been hard on lice. Body lice will infest any kind of cloth, but they do much better on natural fibers, especially wool. Synthetics make it that much harder for lice to prosper. A modern clothing treatment with permethrin is a good way to eliminate lice from a group of people or to prevent infestation if you have to work with lousy people.

Bed bugs and household mites are not likely to be stopped by clothing. Bed bugs bite at night, especially toward dawn, while you are sleeping. You are not likely to be wearing the sort of heavy clothing that will stop bites during those hours. Household mites are so small and active that nothing short of a sealed suit is going to keep them out. About the only reasonable clothing advice in dealing with these pests would be to use light colors on sheets and clothing if you think you have a problem. In that way, you are more likely to notice the bed bug blood spots and the active, tiny mites.

Clothing is a useful tool for the prevention of chigger and tick bites. Both usually climb aboard from the ground or low vegetation and then crawl upward until they reach what they want. The period of crawling on the outside of the clothing, which for them is impenetrable, is filled with danger for these pests. They can easily fall off or get brushed off, leaving them stranded in what may be an inappropriate place. Ticks are conspicuous as they crawl up a trouser leg (especially if the cloth is light-colored), and it is a rare person who would just let the tick go on by without some attempt to capture or kill the pest. Probably the most important action to prevent tick and chigger bites is to stop them from contacting skin above the shoe or sock:

Step 1: Wear long trousers and closed-toe shoes, preferably in light colors that let you spot the ticks easily.

Step 2: Tuck the trousers into socks or boots, or wear puttees or gators. Chiggers and ticks will be forced to take that long, dangerous journey up the outside of your trousers.

Step 3: Spray or rub a good repellent on the openings of the trousers (inner side of the cuffs, waistband, and fly), the exposed parts of the socks, and the tops of the boots. Chiggers and ticks drop off as they pass over the repellent barrier on the way to your skin.

As an alternative or additional measure, you can treat your outer clothing with permethrin (Box 7.1). Given the military experience, permethrin treatment is highly effective against ticks and chiggers, even if you sit on the ground.

For the most part, the air assault bugs are not going to bite through clothing. Kissing bugs bite almost exclusively at night while a person is sleeping. It is hard to imagine one of those large insects completing a blood meal from someone who is awake and alert, though the bugs are surprisingly delicate in their movements. Clothing is largely irrelevant to the prevention of kissing bug bites because they will find a patch of unprotected skin no matter what you are wearing.

Box 7.1 Permethrin-Treated Clothing

Permethrin is a pyrethroid insecticide that has several important characteristics: it lasts well even when it is wet or out in the sun; it is tremendously toxic to insects (and fish, by the way); it repels many kinds of insects at doses lower than it kills them; and it is almost nontoxic to mammals. Permethrin is one of the most common active ingredients in household insecticides.

Back in the late 1930s, the military started experimenting with insecticidal or repellent treatments of cloth. It was concerned with lice that transmit typhus, mosquitoes that transmit malaria and filariasis, and chiggers that transmit scrub typhus. It was soon found that repellents developed for use on the skin were effective much longer on cloth. A series of chemicals was developed, culminating in the mixture called M-1960, which was used until 1974. The smell and irritation of M-1960 finally overcame its usefulness, especially with the repellent DEET so commonly available.

The idea of the usefulness of treated clothing was fixed in the military consciousness by these decades of experience. Permethrin was invented in 1972 as one of a series of pyrethroids that broke the mold for this chemical class by being resistant to environmental degradation. The U.S. Department of Agriculture discovered that permethrin would bind to fibers so tightly that it remained active even after several launderings. Extensive efficacy and safety tests were performed over a period of 15 years, finally resulting in registration for the permethrin treatment of military uniforms in 1991.

Since then, permethrin treatment of clothing has been made available to the public in four different forms. First is an aerosol can that sprays a permethrin solution onto clothing. Once the material dries on the cloth, the treatment remains effective through at least three washings. Second is a solution that mixes with water (emulsifiable concentrate). The mixture is either sprayed onto clothing or used as a dip, usually for the preparation of many items at the same time. Third is the individual dynamic absorption kit, often called the shake-and-bake. A tube with just the right amount of permethrin is emptied in a bag with just the right amount of water and mixed. Clothing is rolled up tightly, placed in the bag, and allowed to wick up the solution. The longer exposure of cloth to the solution seems to result in much longer-lasting protection. Finally, various manufacturing processes integrate permethrin into fibers. These treatments are extremely long lasting, sometimes for years and generally for the life of the garment. You can buy this kind of long-lasting, pretreated clothing in some countries, including the United States.

Spraying insecticide all over your clothing is not exactly what Mother taught you as essential hygiene. Your body definitely absorbs the chemical from the cloth, especially when the garment has been treated recently. Nonetheless, the amounts absorbed are considered safe because they are a small fraction of the harmful doses.

It makes sense to follow label directions to restrict the application to outer garments that don't touch the skin directly (that is, not on hats, underwear, or T-shirts), therefore limiting absorption.

Permethrin-treated clothing is an effective barrier against practically all blood-sucking arthropods. They either die as they contact the cloth or they never penetrate. Treated garments are particularly helpful against the infantry bugs that crawl under clothing (ticks, chiggers, etc.), and they also stop those that bite through clothing. Never forget, though, that permethrin-treated clothing does nothing to protect exposed skin, whether a face, a hand, or a leg.

Biting midges and, especially, black flies crawl about on the skin before biting. They get under the edge of cuffs, under hat bands or long hair, and anywhere the skin is bare. Their mouthparts are short, preventing them from penetrating even the thinnest cloth. Wearing long sleeves, long trousers, and a hat stops many of these annoying flies from biting. Light-colored clothing can make it difficult for these flies to spot you in the first place.

Fleas jump, landing right on your feet, ankles, or legs. Once on skin, they take advantage of their side-to-side flattened bodies and strong legs to move fast between the hairs on your skin. Although most flea bites are on the legs, a flea that lands on your ankle may eventually bite you somewhere much higher up. Long trousers and closed-toe shoes will prevent some bites, as fleas jump on and then jump off without biting, but you will still get some bites if the bottom of the cuff is an open invitation to juicy skin. Even thin clothing stops flea bites, although they are able to insert their narrow sucking tube through thin stockings.

When nature sends in the air force, it pays to be dressed for the occasion. Stable flies, tsetse flies, mosquitoes, horse flies, and snipe flies fly in, land, and start biting. They may bite several times before plugging in, but that just means more welts for us. Perhaps they are in a hurry because their bites are painful, generally resulting in some sort of slapping or swatting defense from us. Horse flies and snipe flies are not as adept at biting through cloth as the others, but all of these pests can usually get through the kind of thin clothing you are likely to wear in warm seasons. They can bite most easily through cloth that is pressed closely to the skin, as usually happens on the thighs, shoulders, and elbows. T-shirts and light knitted sweatpants offer little more protection than naked skin. Woven fabrics pasted to the skin by sweat also present little barrier. The most determined biters can get through blue jeans or, annoyingly, find openings through islets in shoes or clothing.

Some kinds of cloth are very protective. Before the advent of modern repellents in the 1950s, there was much more emphasis on the value of the right kind of clothing to stop bites. Detailed studies in the 1940s resulted in the recognition that herringbone twill weaves and poplin (a square weave made from thicker and thinner threads, giving the textile a ribbed appearance) did not stop mosquitoes. A special cotton weave was developed for uniforms, which was called Grenville cloth in Britain and Byrd cloth in the United States. It was described as a light, windproof twill (a weave that creates a diagonal pattern), and it received generally high marks for comfort in hot climates. Reports of the day showed that Byrd cloth stopped almost all mosquito bites, even in areas with heavy biting pressure. A square weave with nylon threads in one direction and cotton in the other (called nylon-filled poplin or nylon-reinforced oxford cloth) was equally protective. We know a lot less about the protective value of modern textiles, but a few general principles may help you choose the best defense against air force bugs. First, always wear loose-fitting clothing that keeps the probing insect away from your skin over as much of the surface as possible. Second, wear an outer layer if the weather is cool enough. A second layer, like a loose-fitting cotton shirt, not only puts more cloth between you and the biter, it also discourages biting insects by rubbing against the inner layer as you move. Wearing a coarse net undershirt can achieve much the same effect. Finally, take advantage of resistant textiles like tight twills and ripstop nylon while avoiding mosquito-susceptible weaves like lightweight knits.

Do hats help? Mosquitoes rarely bite people through thick head hair, but not all of us have a lot of hair. Bites on the scalp are particularly irritating; therefore, the hair-challenged should certainly wear a hat when out among the biting insects. Even

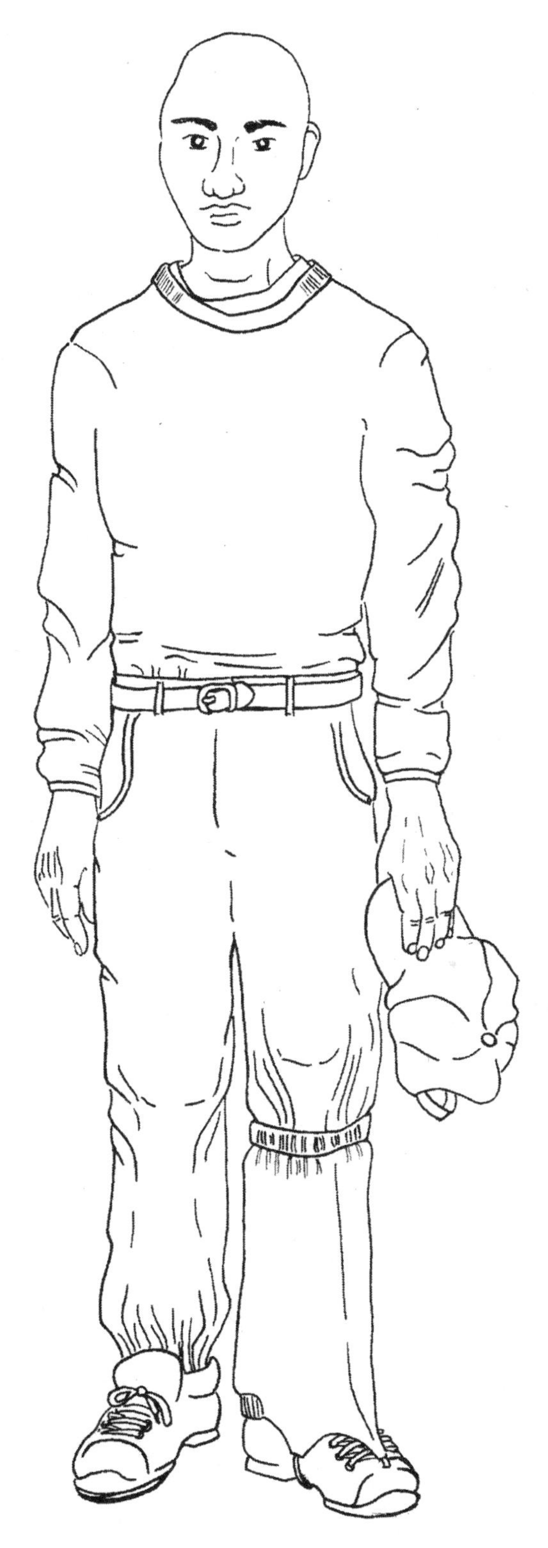

Figure 7.1. Long sleeves, inner shirt, hat, bottom of trousers covered—all this guy needs is some repellent to be completely ready for the bugs.

a person with thick hair will get some benefit by covering the upper forehead with a hat. Horse flies and snipe flies often attack the head or attack from above, so a hat with a brim prevents a lot of annoyance where these flies are abundant.

The ultimate in clothing protection from biting flying insects is net: head net, net jacket, and net pants. Head nets have been around for a long time. Draped from a wide-brimmed hat or sometimes held out from the neck by a ring, they keep the bugs at a distance. The main problem with head nets is that they reduce visibility, especially since they are usually worn during activity outdoors when good visibility is important. One trick to improve vision through a net is to mark the area in front of the eyes with a black marker. The reduction in reflectivity increases visibility. Net jackets are often designed with a net hood and even net mittens. The material needs to be stiff enough to hold itself out from the skin. Net pants are less commonly used, but they are also available. Net jackets with a hood provide nearly absolute protection when biting insects are extremely numerous. It's an odd claustrophobic feeling to be inside your net jacket while hundreds of midges or mosquitoes are a few inches from your face, trying to get in.

We've worked our way down to the clothing we wear and its relationship to biting bugs (Figure 7.1). How much more intimate can we get? Well, we can get topical, as we now turn to topical repellents that change the presentation of our skin to biting arthropods.

PUT ON SOMETHING NATURAL

The most intimate form of protection from biting pests is an application directly to the skin. As a strategy, the application of a product to the skin is a logical choice. After all, the pest is aiming for the skin and the only objective we really care about is preventing that pest from hitting its target. The logic of this strategy has resulted in a series of concoctions that have been developed over millennia. Before the days of modern, or even medieval, chemistry, plants were the principal source of chemicals for all kinds of uses. Plants naturally produce many chemicals that affect insects, and some of them can be used to deter the bugs that bite us as well.

Modern industrialized society attaches a high value to natural products. The public has a long memory for some of the notable disasters created by a few synthetic chemicals that were thought to be safe, but which turned out to be harmful. We went into some detail about DDT in chapter 5, and most people have also heard about PCBs (polychlorinated biphenyls, industrially useful oils that magnify in the food chain and contaminate water supplies) and thalidomide (a supposedly safe tranquilizer that was given to pregnant women by physicians, with disastrous results for the developing child). In contrast, people in a state of nature supposedly lived with natural products and evolved to coexist with them. The reasoning goes that something from a plant must be less dangerous than something synthesized artificially, especially if the chemical does not occur in nature.

REPELLENTS FROM PLANT EXTRACTS

The idea that plant extracts are somehow guaranteed to be safe is ridiculous. Some examples of plant-derived poisons include strychnine from the nux vomica tree, ricin from castor beans, nicotine from tobacco, and poison hemlock. These poisons are famous for their uses in murders, suicides, and assassinations. Of course, many plant extracts are safe enough for almost any application, and we have a long history of experience with their uses. Extracts of lavender, peppermint, and pine find uses in flavorings, scents, and skin products. We are accustomed to having these extracts in products. Their safety is not usually questioned, even though a sufficient quantity would probably be poisonous. Considering that most plant extracts contain many chemicals and that most of those chemicals are not well known, the assumption that they are safe takes a lot of faith. We have that faith in abundance because we have seen our neighbors and families use the products without apparent harm. *Apparent* is a big word here—consider the delayed carcinogenicity of tobacco, which went undetected for hundreds of years.

Seen under the cold, hard eye of technical evaluation, the advantages of plant-based repellent products may seem trivial. However, those advantages are actually important. First and foremost, the public perceives that such products are safer, more wholesome, and less dangerous for the environment. People might even see the application of natural repellents as a positive step in favor of trustworthy chemical products and against infinitely risky, synthetic chemical products. Regardless of the truth of any of these perceptions, they lower the barrier that often exists between leaving a bottle of repellent on the shelf and applying it on the skin. A second advantage is that the aromatic scents and emollient qualities of some plant extracts make these products pleasant to apply. Finally, there is always the hope that a product with a biological origin will be a key that fits more closely into that elusive lock that closes the door on biting pests.

Natural products come in two different forms: extracts that contain a potpourri of chemicals and purified compounds that consist of a single chemical constituent. Extracts are defined only by their botanical sources and the way that they are made. One common method is to heat the plant material in water under pressure followed by distillation of what evaporates. This procedure produces *essential oils*, which are commonly used as flavorings and fragrances. It's a logical step to apply these substances to the skin in order to smell like something other than ourselves to biting pests.

At least 47 different plants have been used at one time or another to produce repellent products designed to be applied to the skin. The first commercial product was sold in 1882 as McKesson's oil. This product was the same as that darling of the natural repellents market, oil of citronella. Isolated from a grass of Indian origin (*Cymbopogon nardus*), oil of citronella always was, and still is, a mediocre repellent product. Its most important active ingredients (citronellol, geraniol, and citral) stop many biting pests but tend to be so volatile that they have little staying power. What is more, some countries classify oil of citronella as a hazardous substance because it can cause skin irritation. Sure smells nice, though.

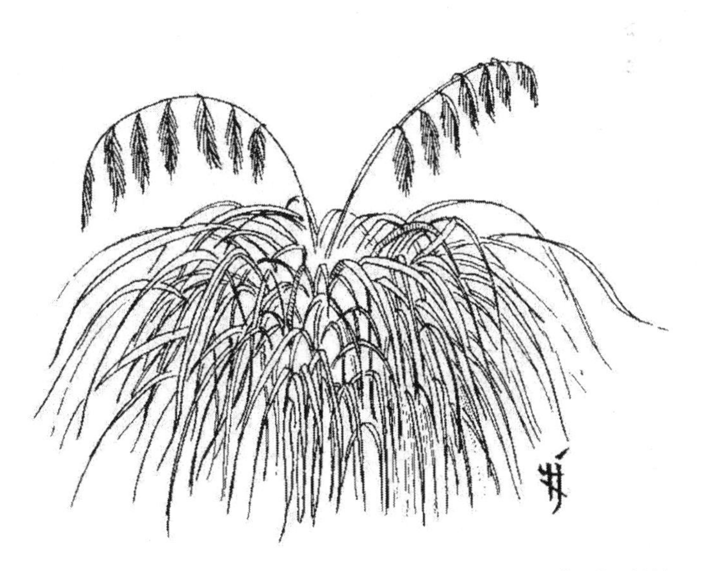

Figure 8.1. Citronella grass (*Cymbopogon nardus*; plant up to 6.5′ or 2 m high).

Other extracts are often mixed together to make fragrant mixtures. Some of these extracts actually repel biting pests, but others do not. The market is filled with botanical mixtures that claim they will protect the user from bites. The active ingredients constantly shift proportions and change constituents, but the claim remains the same. The products state that the lotion, spray, or gel is naturally safe, naturally effective, and naturally pleasant to use. Because the ingredients change so often, because the product claims often sound so blatantly manipulative, and because the standard of evaluation is often suspicious, it is difficult to know whether the attractive bottle with the inevitable picture of a plant actually contains something that will offer real protection from bites.

The selection of a good botanical repellent product is complicated by many factors. First, the huge variety of constantly shifting ingredients used worldwide makes it nearly impossible to provide a comprehensive list of all the plants that might be used. Second, the actual content of a plant extract is going to vary tremendously depending on the method of extraction and the source of the plant. Third, extracts are sensitive to the way the lotion or spray has been prepared, especially because the formulation can slow the release of a volatile chemical that would otherwise evaporate off the skin too quickly. Fourth, the use of plant extracts is only lightly regulated unless pesticidal claims are made, so the safety of the ingredients cannot be taken for granted (more on this later). And, finally, good scientific tests of botanical repellent products and their ingredients are frustratingly rare.

The tables list ingredients that you might see on a botanical repellent label. All of the ingredients have a level of repellency to at least some pests, but their performance varies. As mentioned before, the volatility of a particular chemical can limit its usefulness even if the chemical is highly repellent while it is present. Using oils or other substances can slow the evaporation rate to produce a mixture that gives much longer protection. The potential for a good formulation to convert a less effective ingredient into a more effective ingredient adds considerable uncertainty to recommendations. The ingredients in Table 8.1 have performed poorly in

Table 8.1 Some Plant Extracts with Short Duration of Repellency

Common Name	Scientific Name
absinthe, grand wormwood	*Artemisia absinthum*
andiroba, crabwood	*Carapa guianensis*
anise	*Pimpinella anisum*
basil (holy)	*Ocimum sanctum*
basil (sweet)	*Ocimum basilicum*
bergamot	*Citrus aurantium bergamia*
billy-goat weed, goatweed, Mother Brinkly, blue top, whiteweed, tropic ageratum	*Ageratum conyzoides*
cedar (deodar)	*Cedrus deodara*
cypress (Chinese weeping)	*Cupressus funebris*
eucalyptus (blue), southern blue gum, Victoria blue gum	*Eucalyptus globules* (= *Corymbia globules*)
fennel	*Foeniculum vulgaris*
finger root, Chinese key, Chinese ginger, lesser ginger	*Boesenbergia pandurata*
fishpoison, wild indigo, sarphonk	*Tephrosia purpurea*
garlic (oriental)	*Allium tuberosum*
geranium (rose)	*Pelargonium veniforme*
ginger	*Zingiber officinale*
ginger (cassumunar)	*Zingiber montanum* (= *purpureum*)
lavender	*Lavendula angustifolia*
leech lime	*Citrus hystrix*
lemongrass	*Cymbopogon citratus*
marigold (wild), stinkweed	*Tagetes minuta*
mint (Japanese)	*Mentha arvensis*
mint (peppermint)	*Mentha piperita*
mint (spearmint)	*Mentha spicata*
nutmeg	*Myristica fragrans*
paracress	*Spilanthes acmella*
pepper (black)	*Piper nigrum*
privet (Indian), chaste tree	*Vitex negundo*
rosemary	*Rosemarinus officinalis*
sassafras	*Sassafras albidum*
sesame	*Sesamum indicum*
sweet gale, sweet bayberry	*Myrica gale*
tangerine	*Citrus reticulata*
tarwood, wild pepper tree, teerhout	*Loxostylis alata*
turmeric	*Curcuma longa*
ylang-ylang, kradanga	*Canagium odoratum*

actual tests, though in the right product mixture, they might be quite good. The ingredients in Table 8.2 provided an hour or more protection from at least one kind of biting pest, which is a reasonable minimum standard for a useful repellent product. Table 8.3 lists active ingredients that are either not safe to put on the skin or not safe to use at higher concentrations.

Table 8.2 lists plant extracts that have shown some evidence of providing useful protection from biting insects, especially mosquitoes. Some mixtures have provided hours of protection, appearing to be comparable to synthetic repellent products, but most natural products last much less time. In some situations, and for some people, an hour may be plenty of protection. For example, if you have a quick garden chore and do not want to smell like you just returned from a camping trip, a botanical product that lasts about an hour could be just right.

Table 8.2 Some Plant Extracts with a Strong Contribution to Repellency

Common Name	Scientific Name
andiroba, crabwood	*Carapa procera*
basil (hoary or hairy), *maeng-lak*	*Ocimum canum* (=*americanum*)
betel pepper	*Piper betle*
cassia	*Cinnamomum cassia*
catnip	*Nepeta cataria*
cedar (Alaskan yellow)	*Chamaecyparis nootkatensis*
cedar (western redcedar)	*Thuja plicata*
celery	*Apium graveolens*
chaste tree (roundleaf)	*Vitex rotundifolia*
citronella	*Cymbopogon nardus*
citronella (Java)	*Cymbopogon winterianus*
clove	*Syzygium aromaticum*
coconut oil	*Cocos nucifera*
embrert	*Clerodendrum inerme*
eucalyptus (lemon)	*Corymbia citriodora*
eucalyptus (red), Murray red gum, river red gum	*Eucalyptus camaldulensis*
fever tea, lemon bush, zinziba	*Lippia javanica*
garlic	*Allium sativum*
geranium	*Pelargonium graveolens*
Huon pine, Macquarie pine	*Lagarostrobus franklinii* (formerly *Dacrydium*)
Labrador tea, marsh tea, wild rosemary	*Ledum palustre* (= *Rhododendron tomentosum*)
lemongrass (East Indian)	*Cymbopogon flexuosus*
lily-of-the-valley (European)	*Convallaria majalis*
lime	*Citrus aurantifolia*
litsea	*Litsea cubeba*
makaen, bajarmani	*Zanthoxylum limonella*
Mexican tea, American wormseed, epazote	*Chenopodium ambrosioides*
neem (see Box 6.2)	*Azadirachta indica*
niaouli, cajeput, punktree, tea tree	*Melaleuca alternifolia*
palmarosa, ginger grass, sofia grass	*Cymbopogon martinii*
palm oil, African oil palm	*Elaeis guineensis*
patchouli	*Pogostemon cablin*
pennyroyal	*Mentha pulegium*, *Hedeoma pulegioides*
pine (Scots)	*Pinus sylvestris*
pyrethrum	*Chrysanthemum cinerariifolium*
rue (fringed)	*Ruta chalepensis*
soybean	*Glycine max*
thyme	*Thymus vulgaris*
timur, winged prickly ash	*Zanthoxylum alatum*
turmeric (aromatic)	*Curcuma aromatica*
violet	*Viola odorata*

Figure 8.2. Marsh Labrador tea (wild rosemary) (*Ledum palustre*; plant up to 4′ or 1.2 m high, usually 1.5′ or 50 cm high).

Choosing a botanical extract is complicated by the wide variety of plants that have been used to make repellent products. You might encounter these ingredients anywhere in the world when you pick up a bottle of what you hope is a natural and effective lotion for the prevention of bites. Perhaps a little more detail about the origins, uses, and properties of these plants will help out.

Andiroba, or crabwood, is a tree similar to mahogany and also is used for furniture, veneers, and other applications.

Figure 8.3. European lily-of-the-valley (*Convallaria majalis*; plant up to 1′ or 30 cm high).

Figure 8.4. Litsea leaves and flowers (*Litsea cubeba*; shrub or tree up to 39′ or 12 m high).

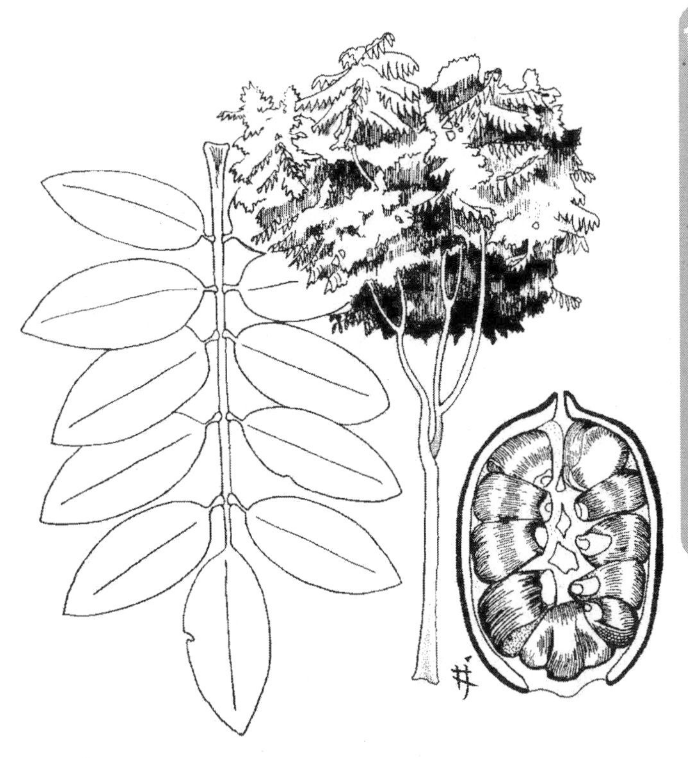

Figure 8.5. Andiroba (crabwood) multileaflet leaf, tree, and seed pod (*Carapa procera*; leaf about 6.5′ or 2 m long, tree up to 55′ or 17 m high).

Like most essential oils, andiroba oil is said to have many medicinal properties. This may be an example of a product that is sold under one common name, but that comes from two different plants with very different effectiveness. A Brazilian study showed that American andiroba oil from one of the crabwood species (*Carapa guianensis*) provided no protection at all from mosquitoes in the laboratory. In marked contrast, a study showed that African andiroba oil from *Carapa procera* protected people for at least an hour against mosquitoes outdoors in the Ivory Coast. For the consumer, it may be a bit difficult to check which andiroba oil you are getting.

Some food items look like they can be quite effective components of repellent products. The basil we use as a flavoring is actually several species of related plants. Fans of Thai cooking know the distinctive flavor of hoary basil, one of the

Figure 8.6. Hoary or hairy basil, one of the species used to make the spice *maeng-lak* (*Ocimum canum*, synonym *americanum*; plant up to 16" or 40 cm high).

species used to make the spice *maeng-lak*. This plant grows as a short herb in gardens, but left to itself in Thailand it becomes a tall tangle of plants (one Thai wag commented that such a tangle was just the thing to flavor elephant rather than the plant's usual use for pork). Other common basil spices do not appear to be very effective as repellents, but hoary basil extract mixed with vanilla provided hours of protection from mosquitoes in one study.

Figure 8.7. Aromatic turmeric (*Curcuma aromatica*; plants up to 3′ or 1 m high).

Turmeric extract (25% in alcohol) was a surprisingly good repellent in one test, providing three and a half hours of protection from three species of pestiferous Asian mosquitoes.

Although not widely tested, celery oil has been shown to repel the yellow fever mosquito in the laboratory for over an hour.

Finally, soybean oil, palm oil, and coconut oil are good carriers for other plant-based repellent ingredients. Each kind of

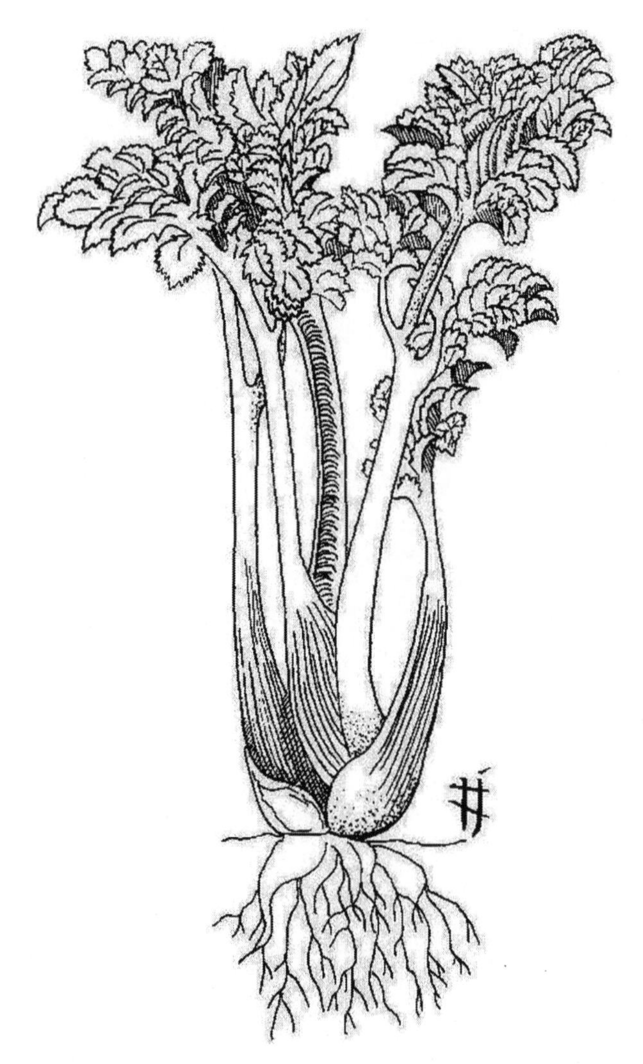

Figure 8.8. Celery (*Apium graveolens*; plant up to about 2′ or 60 cm high).

oil appears to either contribute to repellency or enhance the repellency of other products by mediating the rate at which they evaporate from the skin.

Two of the most controversial ingredients are citronella and lemongrass. These products are extracted from similar grasses that produce nicely scented oils. Lemongrass, of course, is a staple flavoring of Asian cooking. Dozens of products have used these ingredients in repellents, usually at a concentration of 10%. The industry debates the safety of these

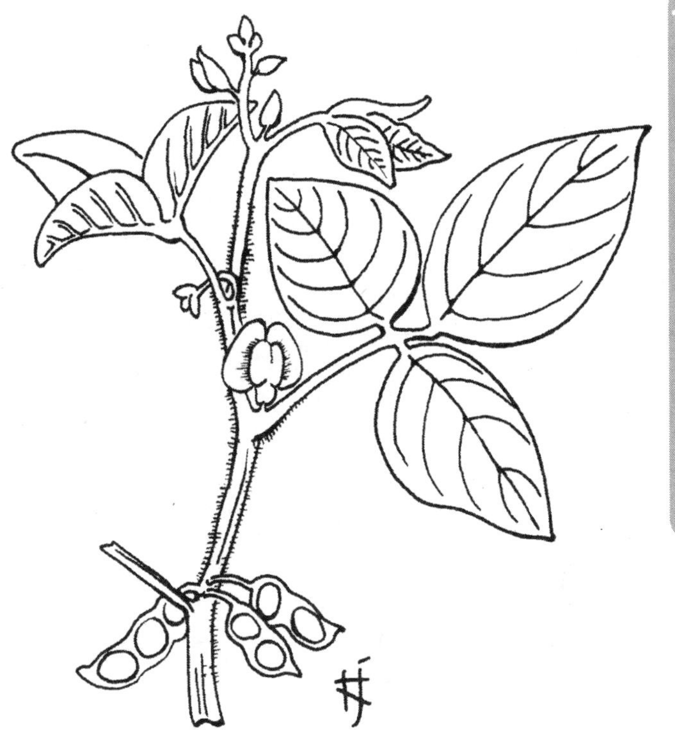

Figure 8.9. Soybean (*Glycine max*; plant variable in height from about 8″ or 20 cm to 6.5′ or 2 m).

preparations. The suggestion to limit concentration to less than 1% is admittedly conservative, based on the maximum content of particular chemicals. Considering the millions of people who have used products with these ingredients, it is a real question whether or not their use at higher percentages is a significant hazard. The hazard from Java citronella may be of more concern because it contains the carcinogen methyl eugenol.

Lemongrass and the citronellas are usually mediocre repellents because the active chemicals are volatile. Unless they are formulated carefully, you can't expect even a full hour of protection. Mixed with the right carriers, some studies have shown protection for two to six hours.

Other plants that we associate more commonly with flavorings than with insect control have been packaged into repellent products at one time or another. It can be difficult to

Figure 8.10. African oil palm (*Elaeis guineensis*; tree up to 65′ or 20 m high).

determine whether the manufacturer wants to improve the product by making it smell better, appeal to our sense that something we eat will be safe on our skin, or actually provide an extract that stops bites. Clove oil certainly repels mosquitoes with its high concentration of eugenol, a chemical that can irritate the skin.

Similarly, thyme contains effectively repellent chemicals that are also sensitizing toxicants.

Figure 8.11. Coconut (*Cocos nucifera*; tree up to 97′ or 30 m high).

Figure 8.12. East Indian lemongrass (*Cymbopogon flexuosus*; plant commonly 3′ or about 1 m high) with Thai bowl of rice.

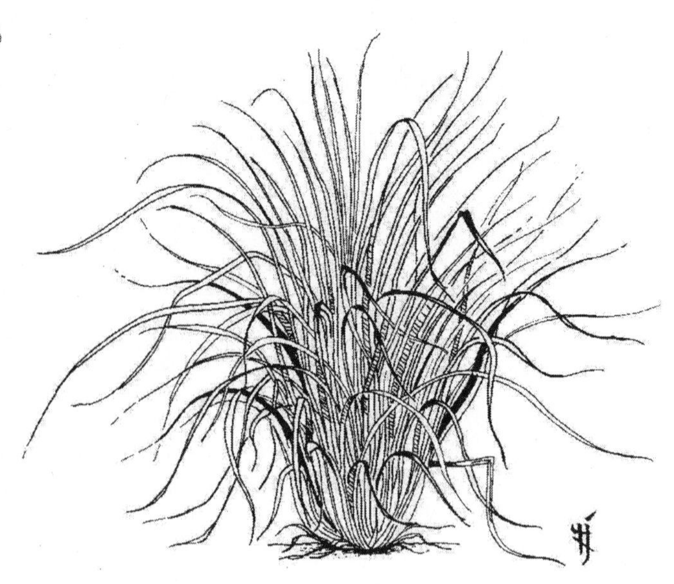

Figure 8.13. Java citronella grass (*Cymbopogon winterianus*; plant up to 5′ or 1.5 m high).

Figure 8.14. Clove stem with florets (*Syzygium aromaticum*; leaf about 3.5″ or 9 cm long, tree up to 65′ or 20 m high).

Figure 8.15. Thyme (*Thymus vulgaris*; plant up to 16″ or 40 cm high).

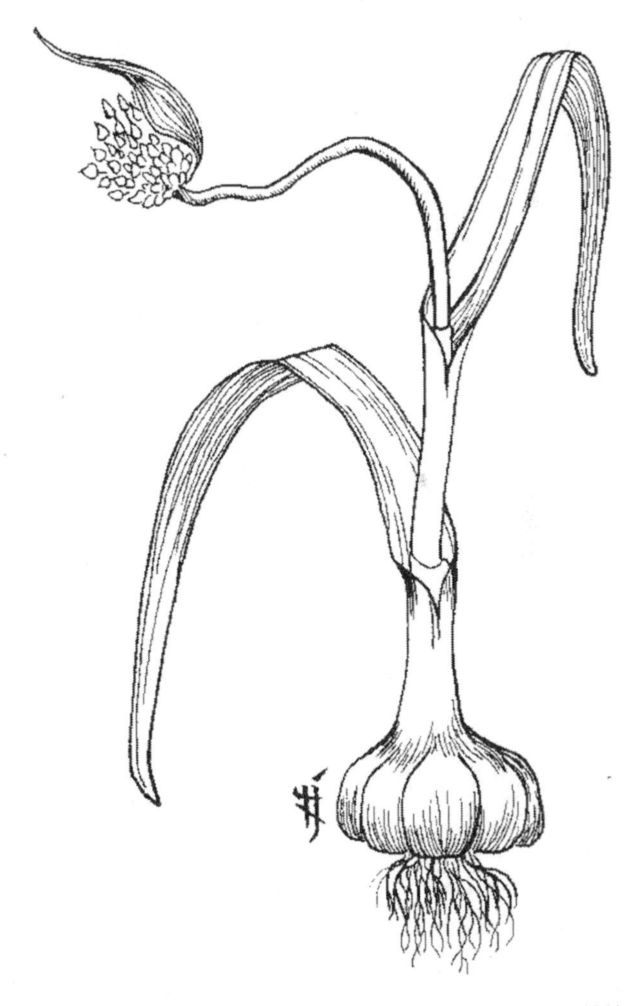

Figure 8.16. Garlic (*Allium sativum*; plant commonly 3′ or about 1 m high).

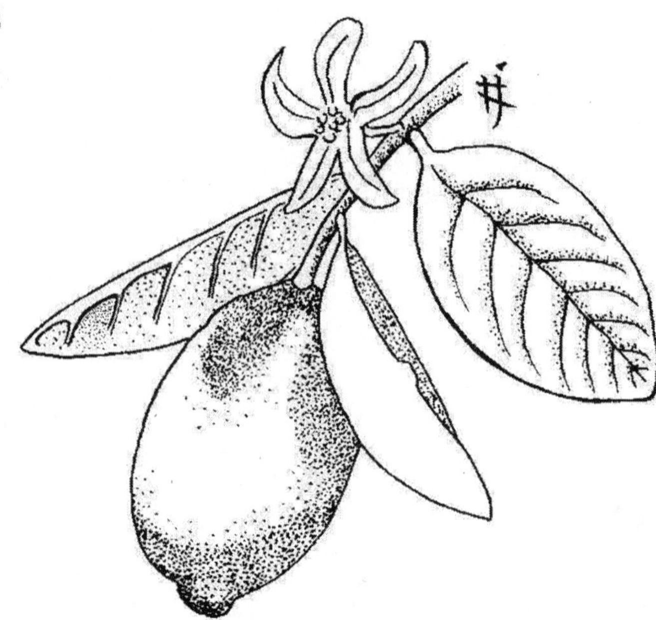

Figure 8.17. Lime stem with fruit (*Citrus aurantifolia*; leaf up to 3.5″ or 9 cm long, fruit up to 2″ or 5 cm in diameter, tree up to 16′ or 5 m high).

Garlic is often proclaimed as a universal repellent, though extracts actually offer only partial protection. The many claims that eating garlic oil will decrease bites seem to be contradicted by scientific studies and by our common experience that people who eat garlic in quantity suffer from mosquitoes as badly as anyone else.

Lime oil (extracted from the peel and not actually eaten often) is a moderately good repellent, but the chemicals in it can be irritating, particularly when combined with exposure to sunlight.

Cinnamon in the form of cassia is from the same hard, woody bark that is sometimes sold as cinnamon sticks. It turns out that this is not the true cinnamon, though it is the most common source of the spice in the United States. The oil was a good repellent in old tests, but it was irritating.

Betel pepper leaves are used as part of a concoction chewed by many people in Southeast Asia. The leaves are smeared with slaked lime (the chemical, not the fruit), wrapped around the fruit of a particular palm tree, and then the whole mess is chewed. Chewing stimulates blood-red-colored salivation, so that a betel

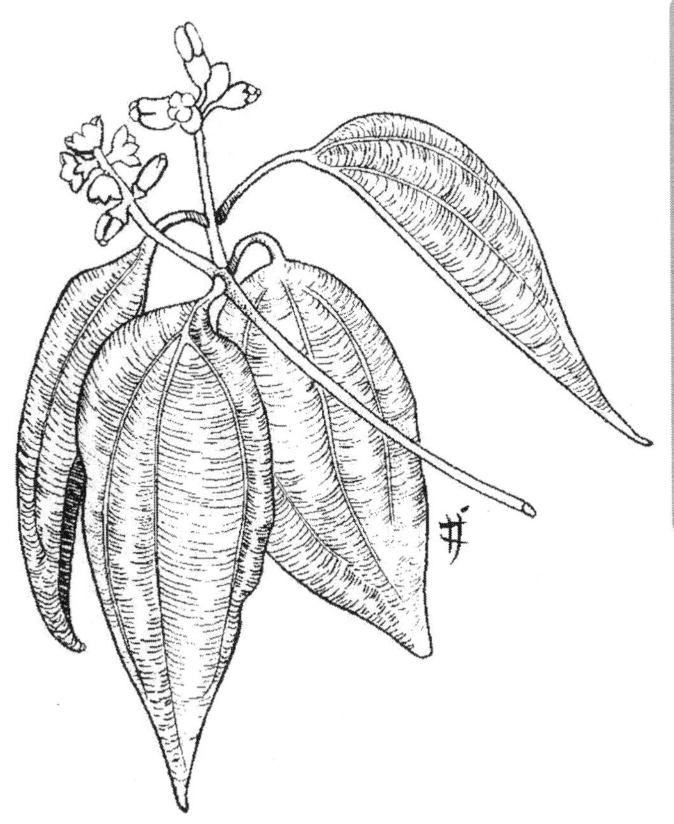

Figure 8.18. Cassia cinnamon stem (*Cinnamomum cassia*; leaf up to 6″ or 15 cm long, tree up to 50′ or 15 m high).

chewer is usually surrounded by pools of red spittle in the dust. The habit is so entrenched in some cultures that it has spawned beautifully worked sets of dishes to hold the various components. As if that weren't enough, people who habitually chew betel find their teeth darkly stained. Why do people take up this unlovely habit? If you ask an old woman in a rural area, she may tell you that it is something to do. If you try the stuff yourself, you'll experience a definite intoxicating stimulation. Perhaps it also keeps the mosquitoes away because the essential oil has been shown to provide protection for about an hour.

The word *cedar* is an imprecise common name designating a variety of coniferous trees. Deodar cedars from the Old World do not produce effective repellent products. Eastern redcedar, though the only source for cedar oil exempted from

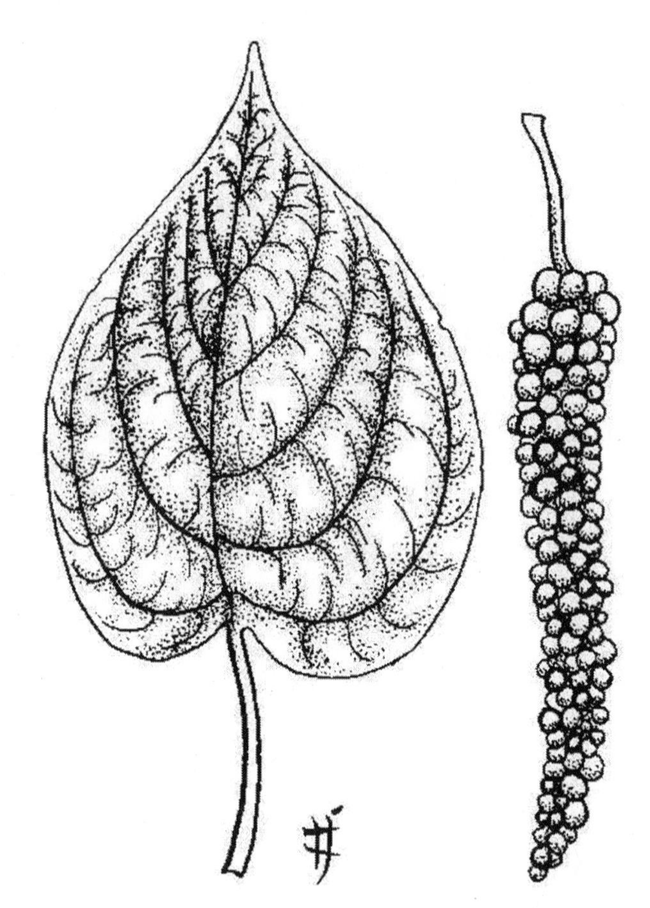

Figure 8.19. Betel pepper leaf and seed pod (*Piper betle*; leaf about 4″ or 10 cm long).

the requirements of U.S. Environmental Protection Agency registration, has not been evaluated thoroughly for repellent properties against biting pests. Western redcedar, on the other hand, produces reasonably repellent oil. Alaskan yellow cedar is a beautiful tree that grows in northwestern North America. Its wood is very resistant to attack from insects and fungus, thanks to the concentration of aromatic chemicals in it. Some groves along the Alaskan coast have suffered a mysterious demise, creating vast acreages of standing dead trees. In part to figure out what to do with these trees, some research has gone into an exploration of the usefulness of oils extracted from the wood. It turns out that one of the major flavor ingredients in grapefruit, nootkatone, is also abundant

Figure 8.20. Western redcedar tree, cone, and leaf (*Thuja plicata*; tree up to 200′ or 62 m high, cone 0.5″ or 1.3 cm long, leaf 0.25″ or 6 mm long).

in Alaskan yellow cedar oil, and this chemical is an effective repellent against ticks. Nootkatone is only safe to put on the skin when it is very pure, which may limit the usefulness of the essential oil.

No word describes eucalyptus like aromatic. At least two species contain chemicals that are highly repellent to biting pests. The red eucalyptus (Murray red gum, river red gum) contains eucamalol, a chemical that repels the Asian tiger mosquito about as well as the standard synthetic repellent, DEET.

The current champion of botanical repellents, PMD (para-menthane-3,8-diol), comes from the lemon eucalyptus. Although there is some confusion in labeling the oil (see below), products that mention lemon eucalyptus as a source of active ingredients are likely to be the most effective botanical repellents currently available.

Figure 8.21. Alaskan yellow cedar (*Chamaecyparis nootkatensis*; tree up to 120′ or 37 m high).

Tea tree oil sounds like an innocuous product from an Australian tree similar to eucalyptus, but various aspects are controversial. In Australia, products are sold on the basis of their antibacterial properties and *Melaleuca alternifolia* is cited as the source. Previously, cajeput oil was considered to come from one species of *Melaleuca*, niaouli oil from another, and tea tree oil from a third. Current taxonomy suggests that the three species are the same, so that when you buy something with one of those names, it is not clear at all where it came from. Another problem is that the content of these oils varies dramatically, including at least one record of an alarmingly high percentage of the carcinogen methyl eugenol. This next fact doesn't have much to do with repellency, but *Melaleuca* is a complete negative in the southern United States, where it is an aggressively invasive species that destroys the ecology of subtropical swampy habitats. Nonetheless, one reliable

Figure 8.22. Red eucalyptus (Murray red gum, river red gum) tree, fruit capsules, and leaf (*Eucalyptus camaldulensis*; tree up to 146′ or 45 m high, each fruit capsule about 0.3″ or 8 mm long, leaf up to 8.7″ or 22 cm long).

record shows that essential oil from *Melaleuca* species is a highly effective repellent against mosquitoes. Even at 1% or 2%, the oil contributes greatly to the ability of a mixture of oils to discourage bites.

Catnip, or catmint, is very attractive to many species of cats. It is a good-looking, small plant with blue flowers. One of the abundant chemicals in the plant, nepetalactone, is important in causing the feline response. The chemical is also a powerful repellent. Efficacy and safety tests on the extracts are few, but concoctions listing catnip as an active ingredient are already on the market.

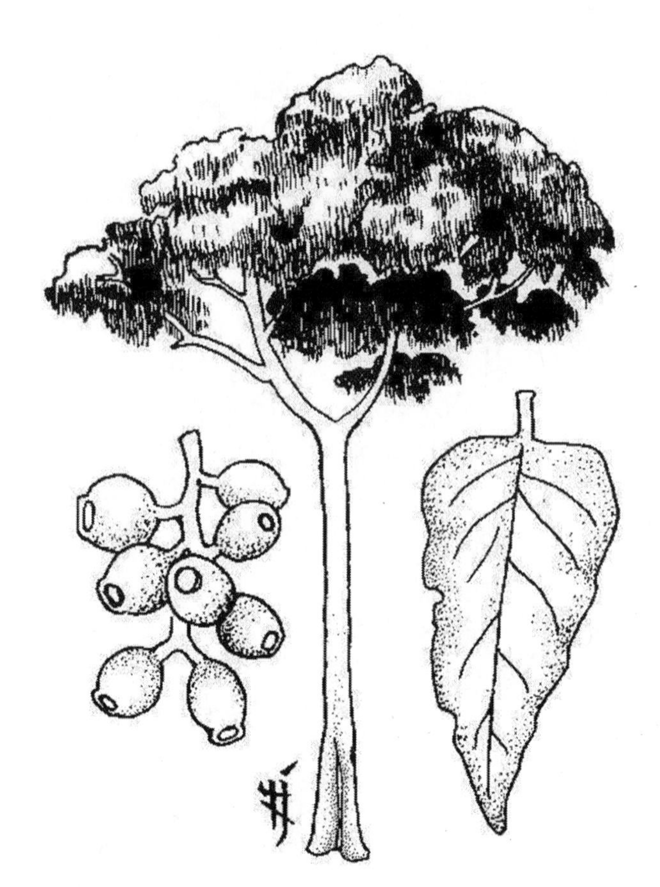

Figure 8.23. Lemon eucalyptus fruit capsules, tree, and leaf (*Corymbia citriodora*; fruit capsule 0.6″ or 15 mm long, tree up to 130′ or 40 m high, leaf up to 8.7″ or 22 cm long).

The round-leaf chaste tree has its own chemical (rotundial) that is a very good repellent ingredient. This plant is related to the common ornamental shrub Indian privet, and it is considered invasive in coastal areas. We do not know much more about rotundial than that it stops mosquitoes biting in the laboratory.

Geranium oil is extracted from the same plant that graces window boxes. It contains a mixture of many chemicals, some of which have repellent qualities. The oil itself is not dramatically effective, stopping the bites of mosquitoes in the laboratory for no more than one and a quarter hours. Although geranium oil does contain some geraniol, an effective repellent, the big source of geraniol is from palmarosa grass. Palmarosa oil is

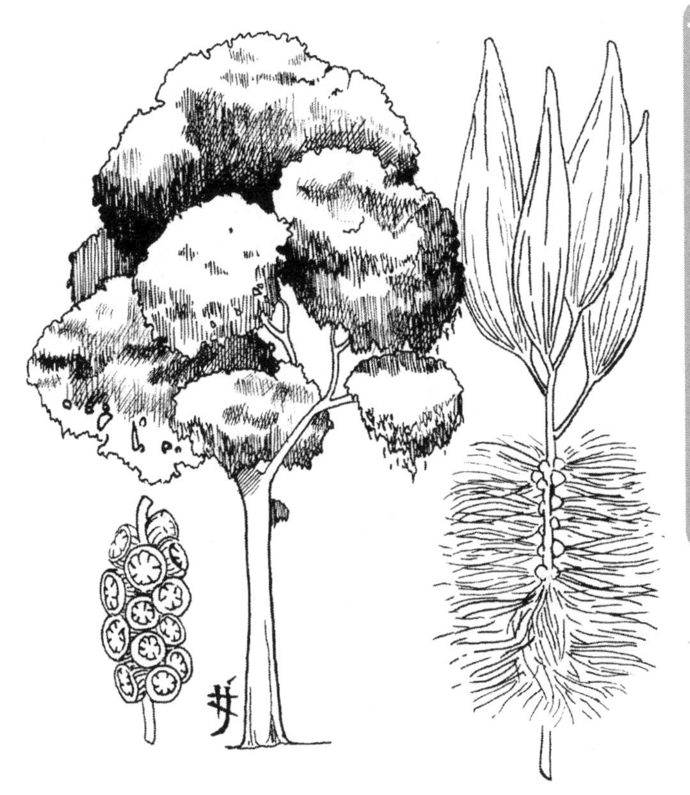

Figure 8.24. Australian tea tree (cajeput, niaouli, punk tree) fruit capsules, tree, and stem (*Melaleuca alternifolia*; fruit capsule 0.1″ or 3 mm in diameter, tree up to 23′ or 7 m high, leaf about 1.4″ or 3.5 cm long).

a pretty effective repellent, but it cannot be used at too high a concentration. There have been products on the market with high concentrations of geraniol that caused skin irritation.

There are a couple of oils that work pretty well as repellents, but they have an important drawback—they are significantly toxic. You can find pennyroyal on the list of active ingredients of some botanical repellents. Alarmingly, one of the compounds in pennyroyal—pulegone—is so toxic that some deaths have occurred from its use. The toxic chemical is produced by two unrelated plants sharing pennyroyal as their common names.

Another toxic active ingredient, sometimes called chenopodium oil, comes from Mexican tea (also called American wormseed or epazote). The plant is a weed that has a wide

Figure 8.25. Catnip (catmint) stem (*Nepeta cataria*; leaf 1.3″ or 3 cm long, plant up to 2′ or 60 cm high).

Figure 8.26. Round-leaf chaste tree stem with flowers (*Vitex rotundifolia*; leaf about 5″ or 13 cm long, bush up to about 3′ or 1 m high).

Figure 8.27. Geranium stem and flowers (*Pelargonium graveolens*; leaf about 3″ or 8 cm long, plant often up to 3′ or 1 m high).

Figure 8.28. Palmarosa (ginger grass, sofia grass) (*Cymbopogon martini*; plant up to about 6′ or 2 m high).

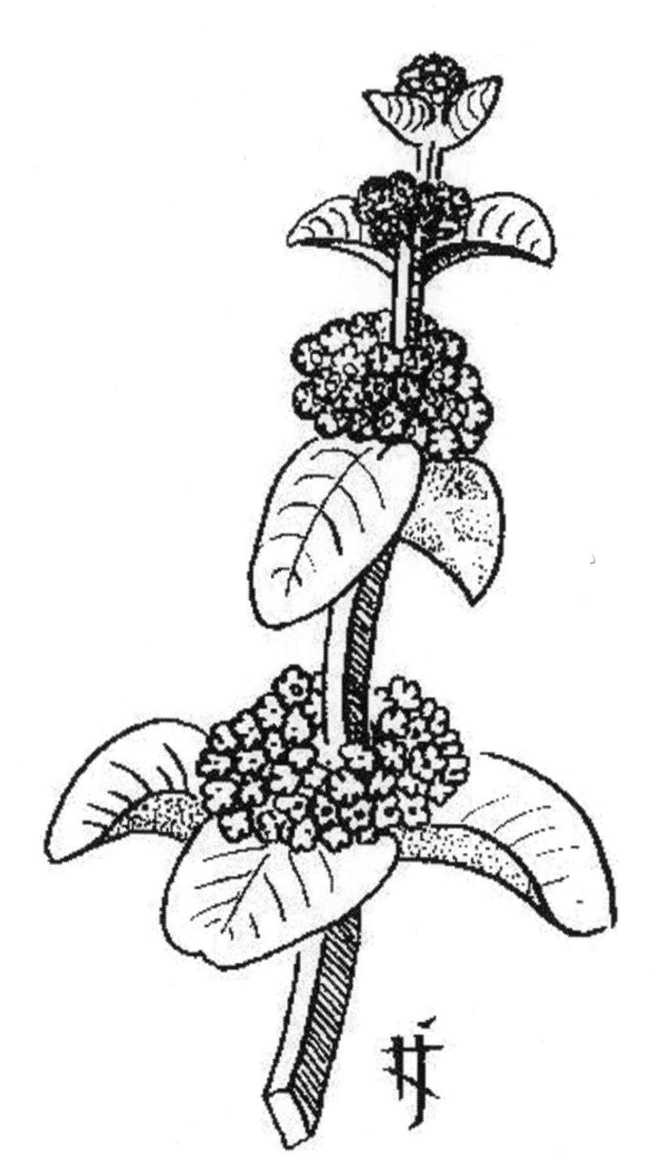

Figure 8.29. Pennyroyal stem (*Mentha pulegium*; plant up to 35″ or 90 cm high, leaf about 1″ or 2.5 cm long).

geographic range in North America, and it is known for its disagreeable odor. The oil has been used to treat intestinal parasites.

Fever tea, or zinziba, is from an African shrub related to verbena. The plant is commonly used by local people as a disinfectant and as a repellent against pests in stored products.

Figure 8.30. Mexican tea (American wormseed, epazote) stem (*Chenopodium ambrosioides*; plant up to 16″ or 40 cm high, leaf about 1.6″ or 4 cm long).

Figure 8.31. Fever tea (lemon bush, zinziba) stem (*Lippia javanica*; plant up to about 6′ or 2 m high).

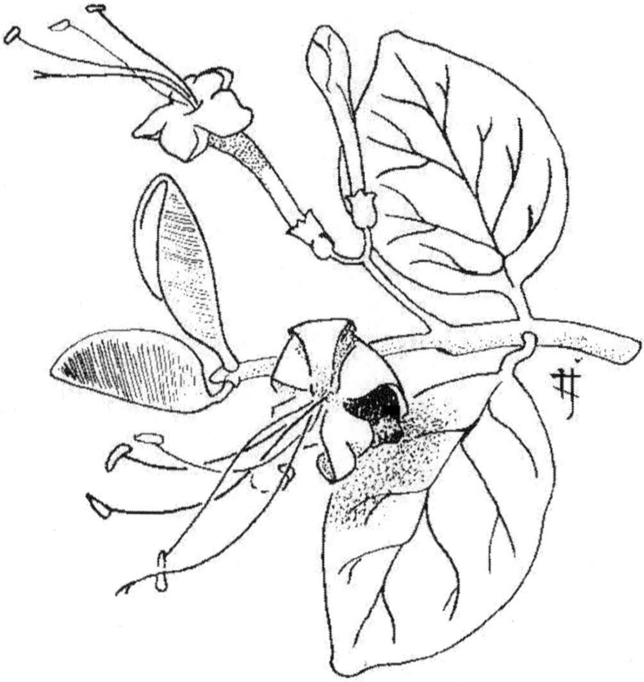

Figure 8.32. Embrert stem (*Clerodendrum inerme*; plant up to 5.8′ or 1.8 m high, leaf about 4″ or 10 cm long).

Figure 8.33. Patchouli stem (*Pogostemon cablin*; plant commonly up to 3′ or 1 m high, leaf about 4.3″ or 11 cm long).

Figure 8.34. Fringed rue stem (*Ruta chalepensis*; plant up to 31″ or 80 cm high, leaf up to 8″ or 20 cm long).

Figure 8.35. Timur (winged prickly ash) stem (*Zanthoxylum alatum*; plant up to 7′ or 4 m high).

The oil contains a large mixture of chemicals, which vary depending on the origin of the plants. One test of the oil showed only moderate repellency.

Embrert is another verbena-like plant that contains a wide variety of bioactive chemicals. The pretty shrub originates from Southeast Asia, though it grows elsewhere as well. It produces an ingredient for Thai cooking and an insecticide that disrupts the normal development of insects. One test of embrert oil as a repellent showed that it provided about three hours of protection from mosquitoes.

They say that, if you lived during the 1960s, you will recognize the aroma of patchouli oil as the common fragrance of social gatherings of the time. It definitely evokes memories of incense-filled dorm rooms and tie-dyed T-shirts. The aroma

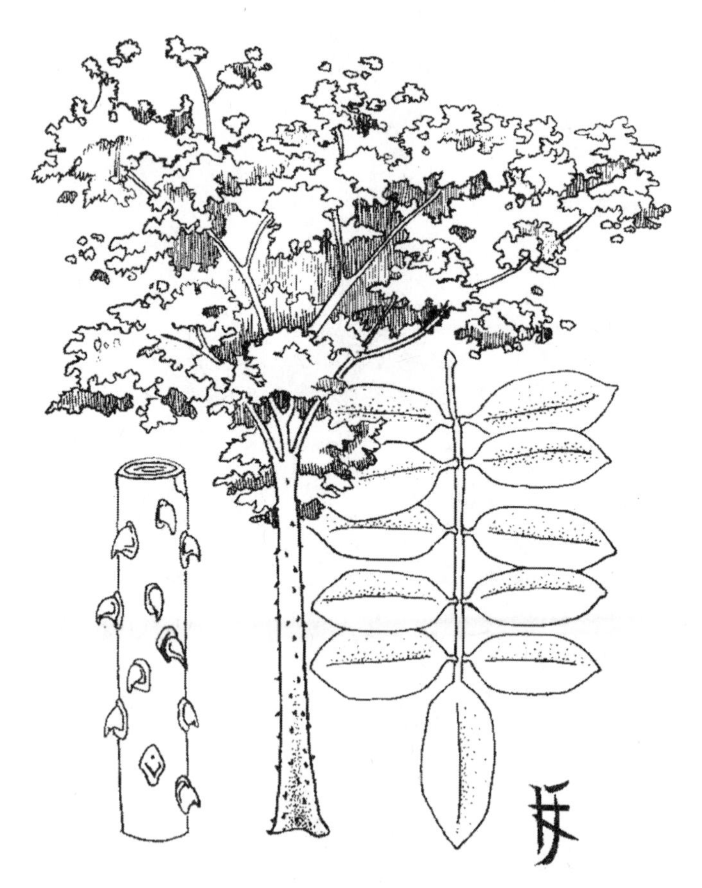

Figure 8.36. Makaen (bajarmani) stem, tree, and leaf with leaflets (*Zanthoxylum limonella*; tree up to 65′ or 20 m high, leaf 14″ or 35 cm long).

has been used to cover up bad smells. Patchouli comes from a shrub that is a native of India and Malaysia. The leaves were used to protect shipments of textiles from India and also to fight against bed bugs. Patchouli oil has considerable repellency to mosquitoes, but pound for pound, it is probably at least 10 times less effective than DEET.

The citrus family includes many species that do not produce the lemons, limes, grapefruits, and oranges familiar to most of us. At least three species from the citrus family have been used as repellents. Fringed rue is a pretty bush with yellow flowers and was one of the plants used in the ancient world as a common, bitter flavoring. Rue is full of folkloric meaning in various countries and is downright toxic if taken orally in excess. Rue oil has some repellent qualities, but it hardly seems worth its toxicity.

Winged prickly ash, or timur, is a commonly used spice in Asia; it is one of the components of the famous "five spice"

Figure 8.37. Scots pine (*Pinus sylvestris*; tree about 60′ or 18.5 m high).

mixture in Chinese cuisine. The essential oil provides good repellency for hours.

The Thai spice plant, makaen or bajarmani, has a limonene-rich oil that has produced a good repellent mixture with clove oil. Tests in Thailand showed that this mixture provided seven hours of protection against a variety of mosquitoes.

Figure 8.38. Huon (Macquarie) pine (*Lagarostrobus franklinii*; tree up to 81′ or 25 m high).

Figure 8.39. Violet (*Viola odorata*; plant about 8″ or 20 cm high).

Pine extracts, including those from the Scots pine, smell nice and have many uses, but they are not impressive as repellents. Huon pine, which is not a true pine, is an exception. The oil from this ancient Tasmanian plant is a highly effective repellent, but it is composed of 97% methyl eugenol—a nice-smelling carcinogen.

Violets are about as innocuous as you can imagine. They are beautiful little plants with purple flowers that are sometimes used in salads. Surprisingly, they contain a wide variety of chemicals and the oil of violets is a very effective repellent. It also contains *trans*-2-hexenal, a chemical that sensitizes the skin and eventually causes a great deal of irritation.

SAFETY OF BOTANICAL EXTRACTS

Safety is a huge consideration in using a botanical repellent, and we make no claim to have reviewed the complete safety profile of the thousands of chemicals present in all of the plant extracts. The EPA regulates insect repellents that, like potions and lotions, are applied directly on the skin. Those products with an EPA registration number have been reviewed thoroughly for safety. In addition, the EPA publishes a "25(b) Minimum Risk Pesticides" (named after Section 25[b] of the Federal Insecticide, Fungicide, and Rodenticide Act, which empowers the Administrator of the EPA to exempt substances from the requirement for registration) list of ingredients that can be used freely without registration, as long as there is no claim of protection from bugs that cause a health problem. What's more, if a product only uses active ingredients on the 25(b) list, then it does not need to be tested for safety as long as the inert ingredients are listed in detail. As a result of this regulation, manufacturers have explored the idea of producing repellents that depend on 25(b) ingredients because such products are exempt from the tens of millions of dollars required to register a new repellent. The list is not without its controversy—for example, Canada and Germany consider oil of citronella to be a hazardous skin irritant whereas the EPA includes it on its 25(b) list of benign ingredients. The fragrance industry would also caution against some of the ingredients based on the potential for them to cause skin problems.

Table 8.3 Ingredients in Botanical Repellents That May Be Hazardous

Common Name	Scientific Name	Maximum Safe Concentration	Hazard
anise	*Pimpinella anisum*	3.6%	based on 0.11% methyl eugenol; carcinogen
basil	*Ocimum* species	0.07%	based on 6% methyl eugenol; carcinogen
bergamot	*Citrus aurantium bergamia*	0.4%	sensitizing and phototoxic skin irritant
cajeput, niaouli, punk, tea tree	*Melaleuca alternifolia*	0.004%	based on 97% methyl eugenol; carcinogen
cedar (Alaskan yellow cedar)	*Chamaecyparis nootkatensis*		likely allergenic contaminants if nootkatone not 98% pure
cassia, bastard cinnamon	*Cinnamomum cassia*	1%	sensitizing skin irritant
citronella	*Cymbopogon nardus*	0.2% or 9%	safety is controversial; based on 0.2% methyl eugenol or 1.3% citral; carcinogen; sensitizing skin irritant
citronella (Java)	*Cymbopogon winterianus*	2%	based on 0.2% methyl eugenol; carcinogen
citrus oils (except others as specified)	*Citrus* species	16%–25%	based on 0.005%–0.0025% bergapten; phototoxic skin irritant
clove	*Syzygium aromaticum*	0.5%	based on 92% eugenol; sensitizing skin irritant
fever tea, lemon bush	*Lippia javanica*	2%	based on 5% citral in related species; sensitizing skin irritant
geranium	*Pelargonium graveolens*	6%	based on 1.5% citral; sensitizing skin irritant
ginger	*Zingiber* species	12%	based on 0.8% citral; sensitizing skin irritant
Huon oil, Macquarie pine	*Lagarostrobus franklinii*	0.004%	based on 98% methyl eugenol; carcinogen
lemongrass	*Cymbopogon citratus*	0.1%	based on 90% citral; sensitizing skin irritant
lime	*Citrus aurantifolia*	0.7%	phototoxic skin irritant
litsea	*Litsea cubeba*	0.1%	based on 78% citral; sensitizing skin irritant
marigold (wild), stinkweed	*Tagetes minuta*	0.01%	phototoxic skin irritant
Mexican tea, American wormseed	*Chenopodium ambrosioides*	prohibited	toxic
mint	*Mentha piperita* and *spicata*	2%	based on 0.1% trans-2-hexenal; sensitizing skin irritant
nutmeg	*Myristica fragrans*	0.4%	based on 1% methyl eugenol; carcinogen
palmarosa, ginger grass, sofia grass	*Cymbopogon martini*	16%	based on 1.2% farnesol; sensitizing skin irritant
pennyroyal	*Mentha pulegium*, *Hedeoma pulegioides*	prohibited	toxic

pine (Scots)	*Pinus sylvestris*	prepare with antioxidants	oxidation creates phototoxic skin irritants
rosemary	*Rosemarinus officinalis*	36%	based on 0.011% methyl eugenol; carcinogen
rue (fringed)	*Ruta chalepensis*	0.15%	based on presence of psoralenes; phototoxic skin irritant
thyme	*Thymus vulgaris*	2%	based on 0.1% *trans*-2-hexenal; sensitizing skin irritant
violet	*Viola odorata*	2%	based on 0.1% *trans*-2-hexenal; sensitizing skin irritant
ylang-ylang, kradanga	*Canagium odoratum*	2%	based on 4% farnesol; sensitizing skin irritant

Despite its limitations, the 25(b) list does imply that a board of toxicologists has thoroughly reviewed the characteristics of each ingredient and decided that it is safe. In other words, you can have more confidence using an ingredient on the list than using one not on the list, though the off-list compound may be perfectly okay. The ingredients on the 25(b) list that are sometimes used in botanical repellents are cedar (eastern redcedar, *Juniperus virginiana*) oil, citronella and citronella oil, garlic and garlic oil, geranium oil, lemongrass oil, peppermint and peppermint oil, soybean oil, and thyme and thyme oil.

Unfortunately, some of the ingredients that have been used in commercial botanical repellents are harmful to the skin. It is difficult to know with certainty whether a particular product will cause irritation, allergy, or a toxic reaction for a particular person. It is also difficult to translate the scientific information on toxicity into firm advice for the person who picks up a bottle with mixed plant extracts and proposes to rub the stuff all over the skin. Table 8.3 attempts to list the clearest examples of repellent ingredients that require some evaluation before they are applied to the skin. We do not claim that the list includes all the potentially harmful ingredients, nor that these ingredients will be a problem for every individual. The International Fragrance Association (IFRA) provided a lot of this information, thanks to the industry's dedicated effort to evaluate the toxicology of ingredients commonly used in fragrances. It is important to remember that the ingredients not included in the table may or may not have been evaluated for safety. The IFRA guidance is based on the use of the active ingredients as fragrances placed on the skin for long periods of time, a use that should be similar to the risks involved with the application of topical repellent products.

Can there be an effective and safe botanical repellent? We think so, but the complications described in this chapter show how difficult the selection can be (Box 8.1). Some of the products on the market provide good protection against a wide variety of biting insects for one or two hours—certainly long enough to get through some outdoor chores or to protect guests while they eat outside. The products change constantly, however, and a repellent with a particular name may change its ingredients from time to time. Many products contain citronella, usually up to 10%. The duration of these products is influenced a great deal by the formulation because the active ingredients are volatile. The right mixture extends the effectiveness by slowing evaporation of the chemicals from the skin. At least one study showed that addition of vanilla to

Box 8.1 Choosing a Natural Repellent Product

Botanical repellents are difficult to evaluate as a class of products. On the one hand, the public badly wants to use what are perceived as natural products. On the other hand, once you use a process to extract everything that's inside a plant, you are getting a mixture of dozens of chemicals. Some plants are not our friends, and many of those chemicals may cause skin irritation or allergy, some cause cancer, and some are downright toxic. If you are like most people and the idea of a plant-based repellent appeals to you, or if you are out where only plant-based repellents are available, how do you make the best decision?

At this point in most of the world, the clear answer is to use products containing the derivatives of oil of lemon eucalyptus with PMD as the active ingredient. It works well and is safe when used according to label directions.

The next layer of selection is much less clear. Citronella has been used by many people, yet even at 10% concentration there is a danger of irritation to the skin. Lemongrass is used similarly, and it has similar potential for causing irritation or allergy. What is more, neither is a long-lasting repellent, though the duration can be improved greatly by correct formulation. Citronella and lemongrass products are not useless, and most are probably safe. We would never recommend them alone as the best protection against a real arthropod-borne disease problem, but they might be just the thing for a short exposure or when you want to offer guests some relief from a problem that is basically a nuisance and not a health threat. The current plant-based products are more likely than the synthetics to be good against one insect and not against another.

Other active ingredients derived directly from plants present more of a problem. As we have described, some are very effective. In fact, some, especially in combination, rival the synthetics in effectiveness. The day may come when we can use such combinations in confidence that one or the other of the ingredients does not include something harmful. In the meantime, we have to face botanical repellents with considerable caution in order to avoid skin irritation or worse.

An alternative is to choose products that use purified extracts or synthesized chemicals that are only part of the raw extract from the plants. In this way, you know exactly what you are putting on your skin. Any problem that results can be evaluated based on evidence rather than on guesswork. Some of those chemicals are effective, and they may provide long-lasting protection if they are formulated with the right oils or other additives to slow down the rate of evaporation.

Whether you use a plant extract, a pleasant mixture of many aromatic plants, or a chemical derived from a plant, be cautious about depending on the product. First, take a good look at whether the product actually protects you from the bites you are trying to avoid. Second and more important, try the product on a small part of your skin to see whether it irritates you. Even if it does not seem to cause any kind of reaction, you may still be sensitizing your skin to damage later on. In the worst case, you may slowly poison yourself with a plant-based toxin. The lists of plants presented in this chapter may be helpful in selecting the right product, but the ingredients change all the time and those who sell plant-based repellents seem to be aggressive about using the latest extract for their new products.

Natural is great, and you can hope that the botanical world will present less of a threat to your long-term health than the world of synthetic chemistry. That said, the principal drawback of botanical repellents is safety and the great unknown of what the dozens of chemicals in an extract are doing to your body. Purified botanical chemicals are safer and, as time goes on, we will know more about them. The impressive efficacy of some mixtures of plant products is an encouraging development that suggests that better products are on their way.

Finally, we suggest that you use a registered product. If there has been some sort of governmental review of the repellent's toxicology and efficacy, you are much more likely to be using a product that is safe and effective.

citronella improved effectiveness dramatically. Another combination sold in several products is soybean and geranium oils, which is apparently effective up to nearly three hours against ticks as well as biting insects. The right extract from lemon eucalyptus can be very effective because it contains PMD, a chemical that rivals any of the synthetic repellents. Combinations of botanical ingredients seem to improve repellency better than combinations of the current synthetic repellents, but the more of those extracts you mix together, the greater the chance that you are going to encounter skin irritants.

REPELLENTS FROM PLANT-DERIVED CHEMICALS

Each of the effective plant-based ingredients prevents biting because it contains chemicals that affect the pests' behavior. Although one or several chemicals may be responsible for most of the activity of the ingredient, an essential oil or other extract usually contains many kinds of molecules, some of which may be in very small proportions. It is not unusual to see scientific articles about new chemicals being found in plant extracts that have been in use for various purposes for many decades. On the one hand, the complexity of botanical extracts is an advantage in that you may actually get some added effectiveness from a compound that no one has discovered or that no one has characterized. On the other hand, the large number of compounds in an extract makes a thorough evaluation of efficacy and safety almost impossible, particularly because the proportions of the chemicals vary between cultivars within a species and between species within a genus and with climate and growing conditions. No one can guarantee that a minor ingredient is not harmful or that there isn't an interaction between some of the many chemicals in the extract.

So far as is known, relatively few of the plant-derived chemicals are important as repellents. Of course, the more that people look, the more chemicals with repellent properties are found, but it is still amazing how such different plants often share the same active ingredient. Some products take the extra step of either purifying plant extracts or synthesizing the same chemicals as those found in plants. This results in a product with much more predictable properties. Let's take a look at some of the chemicals you may find listed on a bottle of a "natural" repellent product.

The evidence for the efficacy of particular plant-derived chemicals is often sparse, but a combination of data and subjectivity can group the compounds in three ways. First are those that provide many hours of protection under the right conditions. Those are linalool, cymene, thymol, carvacrol, terpinene, eucamalol, and PMD. Second, there are a couple of chemicals that provide up to several hours of protection: geraniol and citral. Finally, some chemicals prevent biting for something like one or two hours: eugenol, citronellal, citronellol, pinene, nootkatone, myrcene, limonene (= dipentene), and terpeneol. Most of these molecules occur in more than one form, varying either in the shape of the molecule (the same kind of difference as between *cis* and *trans* fats) or in its handedness (think about the difference between a left- and right-hand glove). The form of the same chemical can make a huge difference in effectiveness, and we understand only some of those relationships. As with the plant extracts themselves, there are also complications with safety. Linalool, limonene, and pinene can oxidize to chemicals that induce irritation, a problem that can be prevented by packaging these compounds with chemical antioxidants. Nootkatone is commonly contaminated with allergens, so it is important to be sure the compound is 98% pure. Skin irritation can result from too high a concentration of citral (more than 0.1%), eugenol (more than 0.5%), or geraniol.

We have mentioned para-menthane-3,8-diol, or PMD, several times in this chapter. The lemon eucalyptus tree (*Corymbia citriodora*) produces this compound in abundance, as well as some of the other chemicals common to repellent plants. It turns out that PMD is a very effective active ingredient, some products providing over five hours of protection against a wide variety of biting pests, with especially good repellency against malaria mosquitoes and stable flies. In 2005, the U.S. Centers for Disease Control and Prevention (CDC) added PMD to the list of repellents recommended for use against the mosquitoes that can carry the West Nile virus. You will see the botanical extract of the lemon eucalyptus called "oil of lemon eucalyptus" or sometimes by its old Chinese name, *Quwenling* ("effective repellent of mosquitoes"). One enterprising company has trademarked this extract as Citriodiol™. In fact, *Quwenling* is not the essential oil officially known as "oil of lemon eucalyptus" but the aqueous product left behind after distilling the oil. *Quwenling* consists of over 60% PMD, so it packs a big repellent punch. Products use dilutions of *Quwenling*, PMD purified from *Quwenling*, PMD synthesized from plant chemicals, or PMD synthesized artificially. Products with just 10% PMD are still useful repellents, stopping the bites of most mosquitoes for one or two hours. The safety of PMD has been better characterized than have some other plant-based repellents, with full registration by government regulatory agencies like the U.S. Environmental Protection Agency and the Canadian Pest Management Regulatory Agency. There are two safety cautions for these products. First, the plant extracts contain citral, which can irritate the skin of some people. Second, PMD itself can cause somewhat more eye irritation than many other active ingredients. The best thing to do is to buy a registered product and follow label directions.

REPELLENTS FROM ANIMALS

Plants are not the only sources of natural repellent compounds. The state of the science is decades behind the botanical world, but twenty-first-century work has begun to reveal many examples of animals that produce substances to protect themselves from many kinds of biting pests. We harvest plant-derived chemicals that presumably protect plants from bugs that eat them. It is something of a biological coincidence that some of the same compounds are also effective against the bugs that bite people. Animal-derived repellent chemicals are different; they are part of a very active defense system against the myriad flies, lice, mites, and ticks that threaten creatures ranging from frogs to humans. Don't forget that those blood-feeding pests are highly dependent on that blood for their survival. Evolutionarily, they will fight hard to overcome those defenses so that a significant number of them will be able to feed, survive, and reproduce. As a result, only a few of the

Figure 1.2. Arizona bark scorpion (*Centruroides exilicauda*; up to 2.8″ or 7 cm long). Photo by Jeffrey B. Knight, Nevada Department of Agriculture.
Figure 1.3. Brazilian yellow scorpion (*Tityus serrulatus*; up to 2.8″ or 7 cm long). Photo by Yuri Messas/Butantan Institute.
Figure 1.4. Omdurman (yellow fat-tailed) scorpion (*Androctonus australis*; commonly up to 4″ or 10 cm long). Courtesy of Giorgio Molisani.
Figure 1.5. Deathstalker scorpion (*Leiurus quinquestriatus*; commonly up to 3.5″ or 9 cm long). Photo by Rittner Oz.
Figure 1.6. Middle Eastern thin-tailed scorpion (*Hemiscorpius lepturus*; up to 3.3″ or 8.5 cm long). Photo by Morteza Johari.
Figure 1.7. South African fat-tailed scorpion (*Parabuthus transvaalicus*; up to 6″ or 15 cm long). Photo by Kelly Swift.
Figure 1.8. Indian red scorpion (*Hottentotta tamulus*; commonly up to 3.5″ or 9 cm long). Photo by Eric Ythier.

Figure 1.9. Brown recluse spider (*Loxosceles reclusa*; body length up to 0.4″ or 1 cm). Photo by David Bowles and Mark Pomerinke.
Figure 1.10. Long-legged sac spider (*Cheiracanthium mildei*; body length about 0.4″ or 1 cm). Photo by Peter DeVries.
Figure 1.11. Black widow spider (*Latrodectus mactans*; body length up to 0.6″ or 1.5 cm). Clemson University/USDA Cooperative Extension Slide Series, Bugwood.org.
Figure 1.12. Banana (armed-banana) spider (*Phoneutria nigriventer*, body length up to 2″ or 5 cm). Photo by Denise M. Candido/Butantan Institute.
Figure 1.13. Sydney funnel-web spider (*Atrax robustus*; body length up to 1.4″ or 3.5 cm). © Department of Medical Entomology, Westmead Hospital, Sydney, Australia, and thanks to Stephen Doggett.

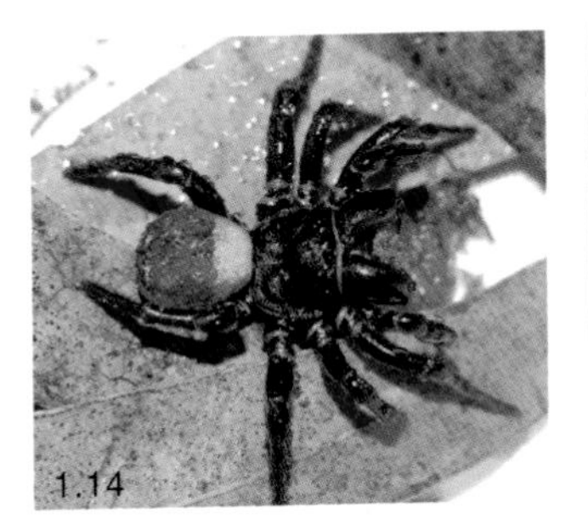

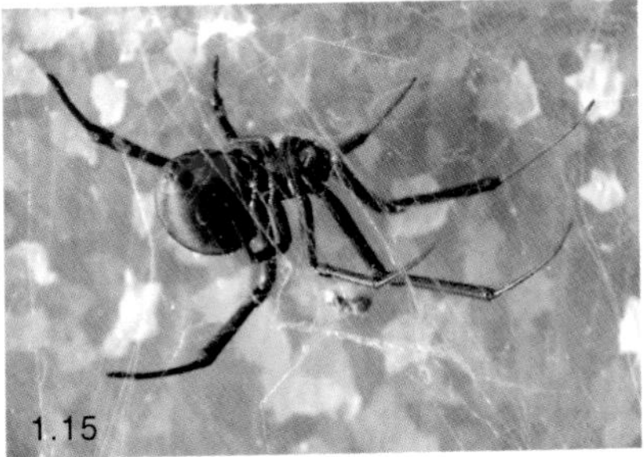

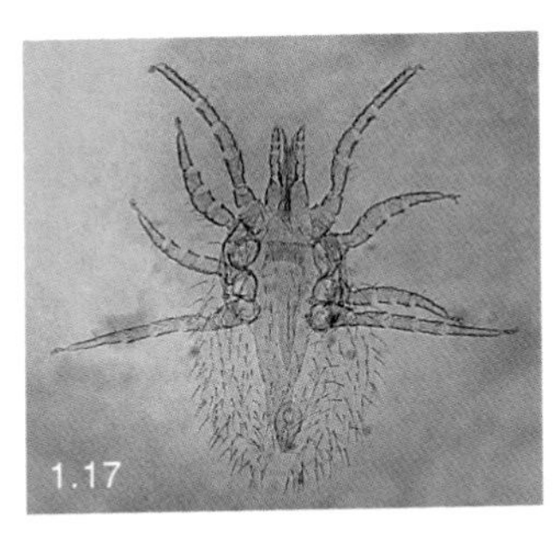

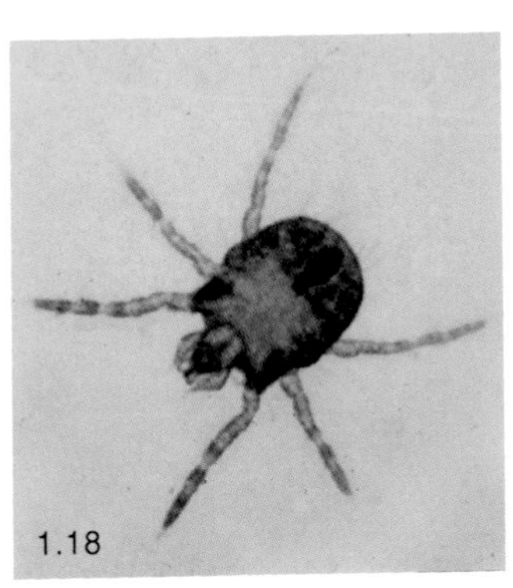

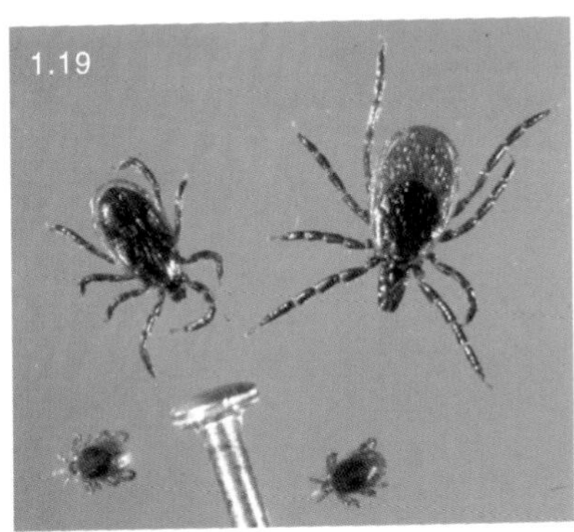

Figure 1.14. Mouse spider (*Missulena* sp.; body length up to 1.4″ or 3.5 cm). © Department of Medical Entomology, Westmead Hospital, Sydney, Australia, and thanks to Stephen Doggett.

Figure 1.15. Australian red-back spider (*Latrodectus hasselti*; body length up to 0.6″ or 1.4 cm). © Department of Medical Entomology, Westmead Hospital, Sydney, Australia, and thanks to Stephen Doggett.

Figure 1.16. Brown widow spider (*Latrodectus geometricus*; body length up to 0.6″ or 1.6 cm). Photo by Fred Santana.

Figure 1.17. Tropical rat mite (*Ornithonyssus bacoti*; up to 0.06″ or 1.4 mm long). © Department of Medical Entomology, Westmead Hospital, Sydney, Australia, and thanks to Stephen Doggett.

Figure 1.18. Chigger (Trombiculidae; larva up to 0.01″ or 0.3 mm long). Photo by Hansell F. Cross, Georgia State University, Bugwood.org.

Figure 1.19. Deer (black-legged) tick (*Ixodes scapularis*; unengorged up to 0.14″ or 3.5 mm long). Photo by Jim Occi, BugPics, Bugwood.org.

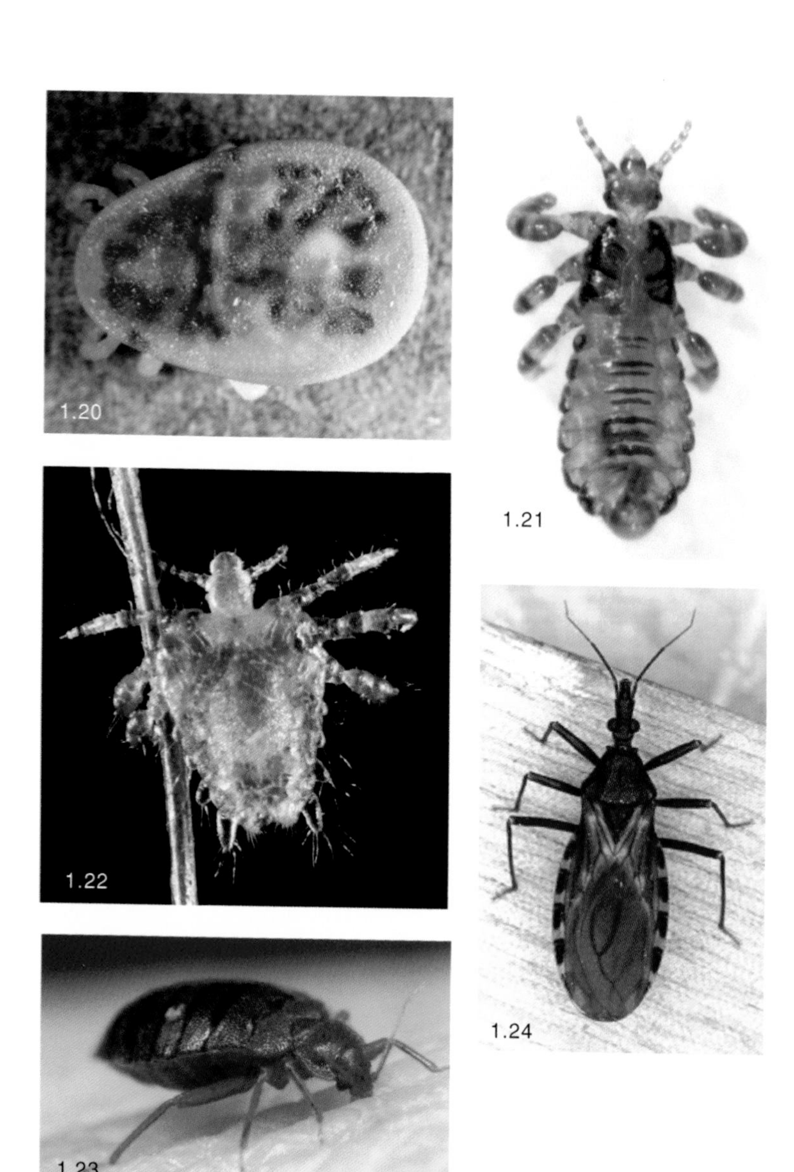

Figure 1.20. Possum soft tick (*Ornithodoros macmillani*; up to about 0.3″ or 8 mm long). © Department of Medical Entomology, Westmead Hospital, Sydney, Australia, and thanks to Stephen Doggett.
Figure 1.21. Head louse (*Pediculus humanus capitis*; up to 0.13″ or 3.3 mm long). © Department of Medical Entomology, Westmead Hospital, Sydney, Australia, and thanks to Stephen Doggett.
Figure 1.22. Crab louse (*Pthirus pubis*; up to 0.08″ or 2 mm long). © Department of Medical Entomology, Westmead Hospital, Sydney, Australia, and thanks to Stephen Doggett.
Figure 1.23. Human bed bug (*Cimex lectularius*; up to 0.24″ or 6 mm long). © Department of Medical Entomology, Westmead Hospital, Sydney, Australia, and thanks to Stephen Doggett.
Figure 1.24. Kissing bug (*Triatoma infestans*; up to 1.1″ or 2.8 cm long). Courtesy of Professor Marcelo de Campos Pereira.

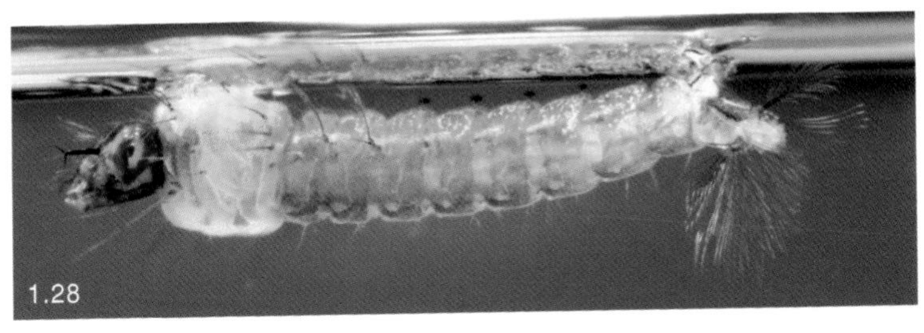

Figure 1.25. Black fly (Simuliidae; up to 0.18″ or 4.5 mm long). Photo by Tom Murray.
Figure 1.26. Sand fly (*Phlebotomus papatasi*; up to 0.1″ or 3 mm long). Photo by Edgar Rowton.
Figure 1.27. Biting midge (*Culicoides* sp.; large species up to 0.1″ or 3 mm long). © Department of Medical Entomology, Westmead Hospital, Sydney, Australia, and thanks to Stephen Doggett.
Figure 1.28. Malaria mosquito larva (*Anopheles annulipes*; up to about 0.3″ or 8 mm long). © Department of Medical Entomology, Westmead Hospital, Sydney, Australia, and thanks to Stephen Doggett.
Figure 1.29. Malaria mosquito (*Anopheles annulipes*; up to about 0.2″ or 5 mm long). © Department of Medical Entomology, Westmead Hospital, Sydney, Australia, and thanks to Stephen Doggett.

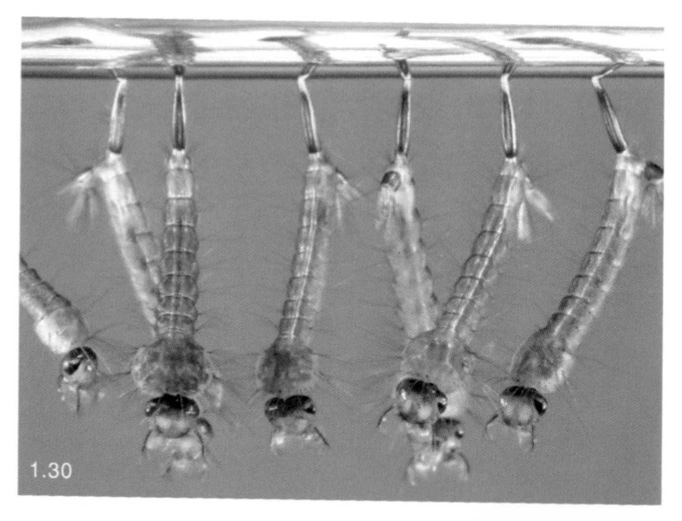

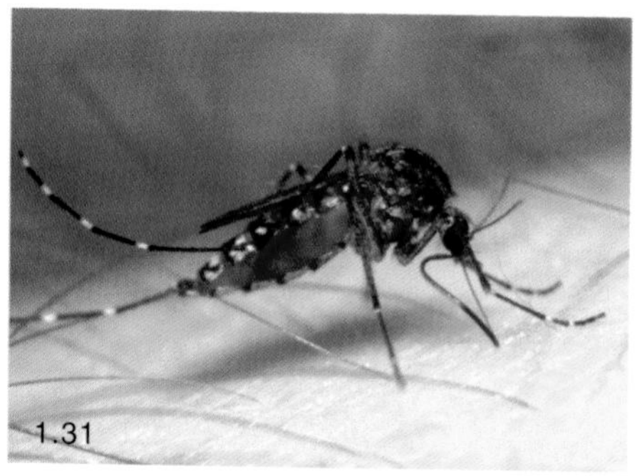

Figure 1.30. Tropical (southern) house mosquito larvae (*Culex quinquefasciatus*; up to about 0.3″ or 8 mm long). © Department of Medical Entomology, Westmead Hospital, Sydney, Australia, and thanks to Stephen Doggett.

Figure 1.31. Floodwater salt marsh mosquito (*Aedes vigilax*; up to 0.24″ or 6 mm long). © Department of Medical Entomology, Westmead Hospital, Sydney, Australia, and thanks to Stephen Doggett.

Figure 1.32. Golden-backed snipe fly (*Chrysopilus thoracicus*; up to 0.35″ or 9 mm long). Courtesy of Dorothy Pugh.

Figure 1.33. Deer fly (*Chrysops vittatus*; up to 0.4″ or 1 cm long). Photo by Sturgis McKeever, Georgia Southern University, Bugwood.org.

Figure 1.34. Striped horse fly (*Tabanus lineola*; up to 0.6″ or 1.5 cm long). Photo by Kevin D. Arvin, Bugwood.org.

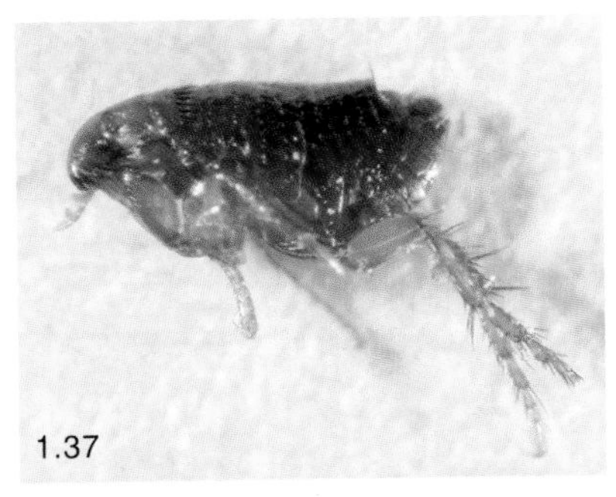

Figure 1.35. Stable fly (*Stomoxys calcitrans*; up to 0.24″ or 6 mm long). © Department of Medical Entomology, Westmead Hospital, Sydney, Australia, and thanks to Stephen Doggett.
Figure 1.36. Tsetse fly (*Glossina fuscipes*; up to 0.4″ or 1.1 cm long). Photo by Steven Mihoc.
Figure 1.37. Cat flea (*Ctenocephalides felis*; about 0.1″ or 2.5 mm long). © Department of Medical Entomology, Westmead Hospital, Sydney, Australia, and thanks to Stephen Doggett.
Figure 1.38. Common yellow jacket wasp (*Vespula vulgaris*; up to 0.55″ or 1.4 cm long). Photo by Tom Wenseleers, Laboratory of Entomology, Zoological Institute, University of Leuven, Belgium.
Figure 1.39. Black and yellow mud dauber wasp (*Sceliphron caementarium*; up to 1.1″ or 2.8 cm long). Photo by Johnny N. Dell, Bugwood.org.
Figure 1.40. Tarantula hawk, a kind of spider wasp (*Pepsis* sp.; size range 0.8″–3″ or 2–7.6 cm). Courtesy of Paul Nylander.

Figure 1.41. Cowkiller velvet ant, a kind of wingless wasp (Mutillidae) (*Dasymutilla occidentalis*; up to 1″ or 2.5 cm long). Photo by Jerry A. Payne, USDA, Agricultural Research Service, Bugwood.org.

Figure 1.42. Bullet ant (*Paraponera clavata*; up to 1″ or 2.5 cm long). Photo by Scott Camazine.

Figure 1.43. Golden northern bumblebee (*Bombus fervidus*; workers up to 0.6″ or 1.5 cm long). Photo by David Cappaert, Michigan State University, Bugwood.org.

Figure 1.44. Honey bee (*Apis mellifera*; up to 0.8″ or 2 cm long). © Department of Medical Entomology, Westmead Hospital, Sydney, Australia, and thanks to Stephen Doggett.

repellent chemicals in animals are generally effective against biting pests. Some animals, like a number of frogs, protect their skin with the same compounds used by plants. A group of tropical birds, the pitohuis, synthesize powerful poisons for their feathers, the poisons similar to those in poison-arrow frogs. Sea birds called auklets emit citrus-smelling chemicals similar to those used by some insects for defense. Perhaps one of the most dramatic examples of a chemical from animals with powerful repellent capability is bovidic acid. This chemical is present in the greasy pelage of the Indian wild cow known as the gaur, an animal that rarely suffers bites from flies or mosquitoes.

For the most part, however, animals emit a complex mixture of volatile chemicals that are characteristic of their species, individual genetics, physical condition, and diet. It may not be too far off to think about the smell of an animal to a mosquito in the same way we perceive the smell of other people. Without getting too detailed about it, we all recognize the smells of our closest family members. To some extent, we can even distinguish their state of health, and certainly their state of cleanliness. Those individual smells contribute greatly to the attractiveness of a particular person to mosquitoes and probably to other biting pests. We only understand a tiny portion of what is going on, but we do know that at least one element of the diet, alcohol, increases a person's susceptibility to mosquitoes. Other dietary effects have been elusive, but they probably exist. A bug searching for some blood is surrounded by smells from many sources, but it orients toward a preferred host in response to the pull of a blend of attractant chemicals and the push of repellent ones. As one animal changes its smell over evolutionary time, a different series of biting pests may respond by going toward the very chemicals that repelled its ancestors.

If you don't mind putting synthetic chemicals on your skin, you can choose from some highly effective products. The synthetics offer many advantages over the natural products, as we'll see in the next chapter.

9 REPELLENTS THAT WORK

The last chapter reviewed natural repellents, mostly of plant origin. The natural repellents constantly shift in composition, make claims that are only partially documented, and mix in a lot of wishful thinking on safety issues. Some botanical repellents, especially mixtures, are effective at preventing bites. They are worth consideration for use, and one active ingredient, PMD, is about as good as any synthetic available.

This chapter moves on to synthetic chemical active ingredients. Synthetics are not necessarily better than natural products, but they usually undergo much more extensive tests of their effectiveness and safety. It is not a surprise that single, defined chemicals are going to give more consistent results than botanical extracts, with all their potential variation from the sources of the plants and preparation of the products.

The list of common synthetic repellent active ingredients is short, but not so short that you won't be faced with real choices at the store. In Table 9.1, we summarize their effectiveness and user acceptability so that you can make an informed choice.

DEET

DEET is an abbreviation for *N,N*-diethyl-*m*-methylbenzamide (also *N,N*-diethyl-*m*-toluamide). You may see these longer chemical names on the product's list of active ingredients. This chemical was first synthesized and developed in a collaborative effort between the U.S. Department of Agriculture and the U.S. Department of Defense in the late 1940s. The chemical has been used by Department of Defense personnel since

Table 9.1 Summary of Important Synthetic Repellent Active Ingredients

Ingredient	Known Effectiveness Against	Advantages	Disadvantages
DEET	chiggers, biting mites, mosquitoes, biting midges, black flies, sand flies, stable flies, horse/deer flies, tsetse flies, fleas	cheap, long safety and evaluation record, broad spectrum protection	oily, distinct odor, melts plastics, irritates eyes, not as effective against ticks, kissing bugs, malaria mosquitoes
picaridin	ticks, chiggers, biting mites, mosquitoes, biting midges, stable flies, fleas	broad spectrum, does not melt plastics, low odor, not as oily, works at lower concentrations	more expensive, less experience with use, not as effective against ticks, some malaria mosquitoes, biting midges
IR3535	ticks, chiggers, biting mites, mosquitoes, biting midges, sand flies, horse/deer flies	extremely safe, long evaluation record, low odor, not oily, does not melt plastics, broad spectrum	repellency sometimes fails at low concentrations
DEPA	ticks, chiggers, bed bugs, biting mites, mosquitoes, biting midges, black flies, sand flies, stable flies, horse/deer flies, fleas	cheap, broad spectrum	oily, distinct odor, melts plastics
PMD	ticks, mosquitoes, biting midges, biting flies	good against malaria mosquitoes and ticks, botanical derivative	only partially evaluated, some preparations have strong odor, irritates the eye
MGK 264, MGK 326	used as additives to DEET products	extends duration increases effectiveness against large biting flies	uncommon in products, registration withdrawn in Canada, controversy over carcinogenicity

Each ingredient may be effective against other pests that have not been evaluated. PMD was discussed in more detail in chapter 8, as a botanical derivative.

1946 and was first released commercially in 1957. It is now the active ingredient in the majority of insect repellent formulations throughout the world. Estimates are that DEET is used by 200 million people throughout the world each year.

DEET is a broad-spectrum repellent, and since its introduction, its effectiveness has been proven against a wide variety of biting pests. Scientific research backs up the claims that DEET provides some protection against mosquitoes, biting midges, stable flies, horse flies, chiggers, ticks, fleas, leeches, and even the tiny larval trematode worms (cercariae) that burrow into the skin from contaminated water to cause schistosomiasis.

This may be a good place to discuss leeches a little. They look like blood-sucking slugs, but they are more accurately described as blood-sucking earthworms. Most species live in the water and only a few of those bite people. Southeast Asia is infamous for its land leeches that inhabit moist forests during the rainy season, where they can be an extremely vexing probiem. You walk along a path for a while, stop for a rest, and see that your ankle has a half dozen black, half-inch leeches sucking your blood. Leeches secrete substances that stop virtually all the pain of the bite (a good thing) and that stop coagulation to prolong bleeding (a bad thing). Fortunately, DEET stops leeches effectively.

DEET's effectiveness against ticks, chiggers, and biting mites depends a lot on the particular species of pest. Labels for DEET include ticks, but the results of scientific studies are decidedly mixed. In one of the rare tests on soft ticks, DEET provided no protection at all. There have been some trials against hard ticks that have shown over two hours of protection, though most studies can only document about an hour of partial reduction from bites. Unfortunately, the tick that transmits Lyme disease appears to be highly tolerant to DEET. On the other hand, chiggers, and by implication biting mites, are even more sensitive than the most sensitive mosquitoes. Lab tests show that low concentrations of DEET kill chiggers in about five minutes. A little DEET could go a long way toward preventing these itchy bites and potential infection with the scrub typhus organism.

We do not usually think of using repellents against lice. Most people will not have a lot of exposure to body lice unless they work in a resettlement camp, prison, or other location that concentrates people under stress. In that sort of situation, it would make a lot of sense to treat your clothing with permethrin. Head lice are a common problem in schools, where the usual procedure is to treat all children's hair with insecticide if even one child has an infestation. The world could use a repellent integrated into hair care products for such a situation, but DEET would not be a good choice because it does a poor job of repelling lice. Presumably, the same would be true of crab lice were someone able to get volunteers for the study.

If you have bed bugs, you need to not have bed bugs, and it is time to do a thorough treatment of the room where you sleep. If you sleep in a questionable hotel, it would be nice to be able to apply a repellent to prevent any chance of bed bug bites. Similarly, it would be comforting to think that you could put a little repellent on your luggage to keep the bed bugs from hitching a ride. Unfortunately, they are not very sensitive to DEET. In two hours, 3 in 20 bed bugs still came for a bite in spite of the application of 75% DEET to the skin. The giant bed bugs (kissing bugs that transmit Chagas disease) appear to be completely tolerant of DEET.

The malaria mosquitoes (genus *Anopheles*) deserve special mention. Travelers often find themselves in a situation where protection from these mosquitoes is not just a matter of comfort, but a matter of life and death. In highly malarious areas, over 15% of the malaria mosquitoes will be able to transmit the pathogen and an exposed person might get dozens of bites a night. The unprotected traveler under those circumstances will almost certainly get malaria. Of course, the risk can be a lot lower in other areas, reducing the odds of getting malaria from a certainty to just a chance. The consequences of infection are so grave that you should do everything possible to prevent bites and to stop the parasite. That means following your physician's recommendations to take preventive drugs during your travels, but it also means using repellents, bed nets, and other measures to reduce the number of bites. The drugs do not make you bulletproof from the threat of malaria and some people have a hard time taking them because of side effects. Malaria mosquitoes are sensitive to DEET, although compared to the house mosquito it takes twice as much DEET to repel the common malaria-transmitting species in India and six times as much to repel the common species in Central America. Field and laboratory trials with good formulations of 20%–30% DEET showed that DEET will last five hours against malaria mosquitoes, but particularly tolerant mosquitoes may be repelled for as little as one hour.

The small biting flies all seem to be as sensitive as mosquitoes to DEET. Studies of black flies in North America showed as much as seven hours of protection, even though those trials used a simple alcohol solution that would have led to the rapid loss of the active ingredient. Similarly, low-percentage (12.5%) DEET in alcohol gave good protection for over two hours against huge populations of biting midges in Utah, with similar protection observed in the American Southeast. One type of sand fly was repelled by less than a third as much DEET as that required to repel the house mosquito. You can expect something better than eight hours of protection from the best DEET-based repellents against those small, irksome, and sometimes dangerous biting flies.

DEET is a mediocre repellent against the larger biting flies. You have to really layer on a good dosage to repel stable flies, and you can expect the relief to last less than two hours. Horse flies sometimes seem to ignore the product entirely or are only repelled for a short time. It might be worth using DEET against tsetse flies; the best DEET repellent available (35% DEET in a polymer base) stopped 91% of bites for an hour and a quarter.

Fleas are discouraged by DEET, though, like bed bugs, you would probably prefer to get rid of the pests entirely. The amount of DEET necessary to stop the rat flea (which carries the bacteria that caused the Black Death) is less than the amount needed to stop house mosquitoes. The sensitivity of the rat flea to DEET means that you could expect long duration protection from good products. Be careful not to apply products designed for humans to your pets: some active ingredients that are safe for us are not safe for dogs and cats.

PICARIDIN

This chemical was added to the list of repellent active ingredients much more recently than was DEET. Its chemical name is 2-(2-hydroxyethyl)-1-piperidinecarboxylic acid 1-methylpropyl ester. Its developmental name was KBR 3023, which you may see in scientific articles. The World Health Organization uses the common name icaridin for the chemical. In the United States, the product is usually called by a registered trade name, Bayrepel, or the common name picaridin. The chemical was carefully developed by Bayer AG in the 1990s using sophisticated chemical modeling methods. It is the first synthetic repellent active ingredient to go through modern efficacy and safety tests as part of its initial commercial development. The product has enjoyed considerable market penetration, with registration in over 50 countries.

Picaridin is commonly formulated at either 10% or 20% concentration. As might be expected, the 10% formulations repel biting pests for a shorter time than the 20% formulations. The active ingredient repels most mosquitoes, though some malaria mosquitoes appear to be more sensitive than others. Significantly, a field trial in Africa showed excellent protection from the principal malaria mosquito for 10 hours following application of 20% picaridin. On the other hand, malaria mosquitoes started biting in only 1 or 2 hours during Australian field studies with a similar concentration of picaridin. Very generally, 20% picaridin provides the same protection from mosquitoes as 30% DEET. You can expect 2–5 hours of protection from most mosquitoes following the application of a product with 10% picaridin and 6–10 hours of protection from a 20% formulation.

A field trial of picaridin against the infamous biting midges of Scotland gave disappointing results. It is not clear whether that particular species was more tolerant, because unpublished trials demonstrated much better results against other biting midges. Similarly, ticks were not repelled well by picaridin in Africa, but unpublished trials showed some level of protection against an important tick in Europe.

Being a fairly new product, there are not nearly as many trials of picaridin's effectiveness as there are for DEET. Enough studies show good efficacy, however, to make a person confident that the active ingredient will work. The problem is that the comparison with DEET may not be a fair one, in that DEET has had so many more chances to fail. From what we know now, picaridin is clearly a highly effective active ingredient similar to DEET in its range of efficacy. Its advantages over DEET are that it works at a slightly lower concentration, it does not melt plastics, it is inherently less greasy, and its odor is less disagreeable.

IR3535

IR3535 is a very interesting repellent active ingredient. Its long chemical name is ethyl butylacetylaminopropionate, sometimes abbreviated as EBAAP. Some consider it to be a natural product because it is derived from a common amino acid building block of proteins, beta-alanine. Merck and Company developed IR3535 as a repellent in the early 1970s by modeling the chemistry of other active ingredients, the first serious attempt to apply molecular modeling to repellent development. IR3535 is a popular active ingredient in over 150 consumer products sold worldwide. The United States has been slow to adopt IR3535. It was not registered until 1999, and only a few American products contain it, although some of them are popular.

IR3535 works about twice as well as DEET against ticks, according to tests performed in Europe and the United States against deer ticks that transmit Lyme disease. Formulations with 20% IR3535 claim eight hours of protection from Lyme disease ticks, but laboratory and field trials have demonstrated something less than complete protection for four hours. It would be nice to have more information, but IR3535 appears to be a better repellent than DEET against ticks. The ingredient is also effective against chiggers and biting mites.

IR3535 repels a wide variety of other biting pests. It is about as good as DEET against the small flies (biting midges, sand flies, and black flies): expect eight hours of protection from the best 20% products. IR3535 is reportedly quite good against the large biting flies, offering about two hours of protection against stable flies, tsetse flies, and horse or deer flies. The chemical also works against lice, but there are no data for bed bugs or kissing bugs.

Mosquitoes are highly sensitive to IR3535. Formulation makes a big difference in the duration of effectiveness, as is the case for all active ingredients. One encouraging sign about IR3535 is that some tests with simple alcohol solutions have resulted in over 10 hours of protection, suggesting that better formulations would be highly effective. Applications of the best 20% formulations prevent bites from house mosquitoes and their relatives for 5–13 hours. Yellow fever mosquitoes and other species in that genus are about equally sensitive, repelled 5–10 hours by a 20% formulation. Extensive tests of the important African malaria mosquitoes show that IR3535 can commonly provide 6 hours, and as many as 10 hours, of protection. Some trials with 20% IR3535 have failed to show such good results, repelling malaria mosquitoes for only 2 or 3 hours. Important Asian malaria mosquitoes were repelled for 4–5 hours. The bottom line is that a good 20% formulation of IR3535 should protect you from most mosquitoes for 6–10 hours and from malaria mosquitoes for 4–6 hours.

The professional community often represents IR3535 as 10%–20% less effective than DEET, but the numbers do not seem to bear this out. Good formulations of the active ingredient are at least equivalent to DEET and actually better against large flies and ticks. The chief advantages of IR3535 over DEET are its almost perfect safety record and its low odor. Once the best formulations of IR3535 dry on your skin, it's hard to tell that they are there—except that the biting stops.

DEPA

You won't find DEPA in many places outside India, but in that important country DEPA has become one of the principal repellent active ingredients. It was developed in the early 1980s to provide a lower-cost alternative to DEET. The problem was that one of the chemicals used to make DEET was expensive in India. By making a very similar molecule (*N,N*-diethyl-phenylacetamide—actually discovered in the same program that developed DEET), researchers were able to avoid the expensive starting material. The result has been a repellent that is similar to DEET in efficacy but that costs 86% less to manufacture.

Comparisons between DEET and DEPA have shown remarkably similar effectiveness. Soft ticks, hard ticks, bed bugs, house mosquitoes, a malaria mosquito, sand flies, black flies, horse flies, stable flies, and fleas were all sensitive to DEPA during trials conducted in India. The tests suggested that DEET, as well as DEPA, is effective against soft ticks, horse flies, and stable flies, which is in contrast to many studies in other parts of the world. This is an example of how evaluations, particularly field evaluations, of repellent products are challenging to perform in such a way that the results can be compared. The particular species tested, the conditions of the test, and the methods used for the test can all make big differences in how well an active ingredient or repellent product performs (see chapter 11).

OLDER ACTIVE INGREDIENTS AND ADDITIVES

There are some older active ingredients you may find in repellents, especially if you go overseas. Two that appear in a few of the American registered repellents are MGK 264 (*N*-octyl-bicycloheptene dicarboximide) and MGK 326 (di-*N*-propyl isocinchomeronate). A few American products contain 25% DEET, 5% MGK 264, and 2.5% MGK 326. These are the highest concentrations of the two MGK chemicals allowed by the current U.S. Environmental Protection Agency's registration. The addition of these ingredients improves the performance of DEET by slowing its evaporation. They also fill in the gaps of DEET's repellency by doing a better job of deterring the bites of stable flies, horse flies, and probably malaria mosquitoes. By some estimates, the addition of MGK 264 and MGK 326 extends the duration of DEET significantly. One way to think of it is that these additives can transform a lower-percentage DEET product into one that performs as well as a higher-percentage product.

Dimethyl phthalate (DMP, sometimes DIMP) and ethyl hexanediol (EH) were the two main active ingredients in repellents before the advent of DEET. China and India used DMP extensively until it was replaced by lemon eucalyptus products and DEPA. EH, sometimes called Rutgers 612, was still used as the U.S. Air Force's survival vest repellent in the 1980s. You may find products with DMP or EH when you are traveling and wonder whether they could work. They are both broad-spectrum repellents, generally about half as effective as DEET. They may actually outperform DEET against tsetse flies and are similar to DEET against black flies and sand flies. Back in the day, DMP and EH were mixed with one or two other active ingredients in order to achieve longer-lasting, broader-spectrum repellency. The formula "6-2-2," or M-250, consisted of six parts DMP, two parts EH, and two parts indalone. Another mixture, M-2020, had four parts DMP, three parts EH, and three parts dimethyl carbate. Mixtures are still available in some countries, and they actually do make the products more effective.

HOW TO USE REPELLENTS

A third of the U.S. population uses a product containing DEET at least once a year—and that does not count people who use repellents with other active ingredients. It is hard to imagine meeting someone who has no concept of a material that can be spread on the skin to stop bites, whether they call it a repellent, bug juice, jungle juice, or mosquito dope. Yet, if you go far enough, you can find a few folks who have never experienced an effective repellent. Give a person like that a bottle of 75% DEET in the middle of a mosquito-infested swamp, and you will first see a look of complete astonishment followed by a broad smile. There is no question that repellents can make the difference between itchy misery and comfort.

There is more question about the ability of repellents to protect people from malaria, dengue, and other pest-transmitted pathogens that cause disease. Common sense would tell you that the fewer bites, the less chance you have of getting one of these illnesses. That is true whenever the proportion of infected pests is low, but not so true when the proportion of infected pests is high. Repellents are seldom 100% effective because certain individual biting pests are more tolerant than the population as a whole; because imperfect application leaves parts of the skin unprotected; and especially because abrasion, sweating, and wetting tend to lower the concentration of the active ingredients. What is more, when we say that a repellent product will last two, six, or eight hours, we are not taking into account the activities of the individual, the impact of extreme weather conditions on effectiveness, nor the particular population of pests. Even if you use repellents, you are likely to get at least a few bites, and there is no guarantee that you will be lucky enough to avoid the bites of the infected pests. Several repellents have been shown to reduce the number of cases of particular arthropod-associated diseases, but they have never been shown to eliminate the risk entirely.

Perhaps the first step toward using repellents as effectively as possible is to decide why you want to use one. First, what is the pest? A common theme throughout this book has been the importance of knowing what is biting you. The guidance for identifying pests in chapter 1 is more than an interesting piece of natural science—it really makes a difference. Second, who is going to be using the product? There is specific guidance for the use of repellents on children, but other people may have particular sensitivities as well. Third, what activity is involved? Just from a practical standpoint, your choice of products and their use will depend a lot on whether you are having a picnic, fighting a war, or attending a business convention. And fourth, what is the threat from the bites? Bug bites can vary from a minor irritation, to a dangerous allergy problem, to the source of a fatal infection. We will now answer these questions in detail.

Box 9.1 Considerations on Using Repellents

- Use a registered product.
- Follow label directions.
- Apply as much as you need to exposed skin and outer clothing.
- Wash off the repellent when you are no longer exposed to biting pests.
- Wash repellent-treated clothing after you no longer need the protection.
- Do not allow young children to apply their own repellent.
- Keep repellents out of your eyes, mouth, nostrils, ears, and other orifices.
- Do not spray aerosol repellents directly on your face; use your hands.
- Apply repellent far enough from your eyes so that sweat does not wash it into them.
- Try to avoid treating the palms of your hands, or wipe them off after application.
- If you apply repellent over sunscreen or sunscreen over repellent, be alert to a decrease in protection from both products.
- Do not apply to irritated, sunburned, or damaged skin.
- Expect variation in performance. Try different products, and match the product to the pest and to the conditions.
- Never use a repellent for humans on an animal and *never* use a repellent designed for animals (including collars) on a human.
- Get a product that is comfortable. A repellent that remains in the bottle is always ineffective.

WHAT IS THE PEST?

Biting mites can be a vexatious problem in a house, one that requires some heavy-duty pest control action. No one would recommend using repellents as your first line of defense against these pests, but there are at least a couple of situations when it would be nice to be able to rely on a product for protection. Typically, a person unfamiliar with biting mites suffers with the problem for some time before even noticing the tiny creatures. Then there is usually a delay while the person finds out what they are and what to do about them. As we reviewed in chapter 4, you want to stop them at their sources by eliminating rodents or nests in your house or by treating your pets. These measures take time to apply and time to take effect. In the meantime, a little use of repellents can prevent a lot of bites from mites. You will probably want a long-lasting formulation of DEET, DEPA, picaridin, or IR3535, since most people spend many hours in their homes. The best formulations last eight hours or longer, usually with 30% or greater DEET or with 20% picaridin or IR3535. The picaridin and IR3535 products will have a real advantage in this situation because they have less odor, better skin-feel, and do not melt plastics and varnishes. The mites can only get on you by crawling. This habit gives you the advantage because you can use the repellent as a barrier on your skin between where the mites are coming from and where they might bite (often, where skin touches skin). Try also applying the repellent to your clothing, especially where there are openings like cuffs or sleeves. The other situation where repellents might be the right defense against biting mites is when you visit a place that is infested. Of course, you probably won't know until you've gotten some bites, but at least you will be able to limit the damage.

Chiggers are mites, but they attack people outdoors, getting on board directly from the ground or from the tips of vegetation. They are especially fond of biting either where skin touches skin or where clothing is tight. They often crawl from your shoes up your legs to finally lodge around your waistband, in folds of skin, or in your groin area. The same repellent active ingredients are effective, but in this case the disadvantages of DEPA and DEET (oily, more odor, melt plastics) compared to picaridin and IR3535 may not be important, since you would presumably be involved in some vigorous outdoor activity where odor and skin-feel are less important. Permethrin-treated outer clothing (see chapter 7) can do a world of good, but you can achieve much the same effect temporarily by applying repellent to your clothing. In general, the repellent will last on the clothing for at least a day and probably longer, until the items are washed. Special attention to the treatment of the cuffs, fly, and waistband of trousers can be as effective as treatment of the entire trousers. A little application to the tops of your boots, the exposed portions of your socks, and your lower leg will help to protect the rest of your body. Treatment low down is especially effective if you are careful not to sit on the ground where chiggers might be present. If you are in a part of Asia, Australia, or the Pacific islands where there is scrub typhus, you will want to be extra careful to avoid chigger bites. If you are going to be hiking or working outdoors, be sure to wear long trousers and to treat them with permethrin.

Ticks are a real problem. DEET, DEPA, picaridin, and IR3535 all claim some effectiveness against them, but we know from experience and quantitative trials that protection varies. IR3535 probably has the best potential, based on the much greater sensitivity of ticks in laboratory tests. The strategy for application is much like that for chiggers, though ticks are not always so oriented toward places in skin folds or between skin and cloth. Permethrin treatment of your trousers is probably even more effective against ticks than chiggers. There are not many places in the world where there are no tick-borne viruses, bacteria, or parasites. The risk is reduced somewhat by the general rule (with plenty of exceptions) that a tick attached for less than 12 hours has virtually no chance of transmitting its pathogen. As a result, the careful removal of a tick soon after attachment greatly reduces the chances of getting one of those nasty infections. If you are hiking or working where ticks are abundant, the prudent course is to use permethrin-treated trousers tucked into boots and to apply IR3535 to the upper parts of the boots, exposed portions of the socks, and tops of the boots. Lice, bed bugs, and kissing bugs are not particularly sensitive to most repellent products. DEET has some effect on lice, which causes us to assume that picaridin and DEPA probably also repel lice for a period of time. The use of repellents against lice is not a standard practice, making it difficult to recommend any particular product or procedure. IR3535 has been used successfully to prevent infestations with head lice, which represents the state of the art in this area. For the time being, the best advice is to rely on registered louse treatments (pediculicides) and use them the old-fashioned way. In contrast, we could really use a good bed bug repellent; it's just that we don't have one. In one trial, DEPA and DEET provided up to eight hours of protection against the tropical species of bed bug. That would be adequate protection for someone accustomed to a short night's sleep and certainly better than nothing if you were forced to sleep in a buggy hotel room. But permethrin-based bed bug treatments are a better option, as described in chapter 5. Kissing bugs are the worst. They can transmit a dangerous and often incurable pathogen, besides the fact that they fill up with a disgusting amount of your blood. Most unfortunately, we do not have a repellent that reliably stops kissing bugs from biting. Use your bed net carefully in rural Latin America.

The vast majority of repellent applications are for protection against mosquitoes. We have seen that the house mosquito types (genus *Culex*) tend to be the most sensitive, with the yellow fever mosquito types (genus *Aedes*) not far behind. The malaria mosquitoes (genus *Anopheles*) are the most tolerant, but all of the major repellent active ingredients are usefully effective. In fact, all of the active ingredients discussed in this chapter, even the older ones that are less effective, provide something close to complete protection against mosquitoes immediately after application. The usual measurement of comparison between products is the length of time that the repellent continues to stop the bites. A clean comparison between a number of products containing 10% or 20% of the active ingredient only tells part of the story because some active ingredients are used at higher concentrations. DEET, because of its long history, is commonly sold as the 98%–100% technical chemical, even though a good formulation of 35% DEET protects a person just as long. Manufacturers and regulators have been cautious with more recent active ingredients, generally restricting picaridin to no more than 20%, IR3535 to no more than 30% (20% in the United States), and PMD to no more than 10% (26% as a botanical extract).

Unlike mites, chiggers, and ticks, mosquitoes fly right to the area of your body where they will bite. As a result, you have to apply the repellent to every square inch of exposed skin that you want to protect. Given the quantity of repellent required, especially if your legs and arms are not covered by clothing, it is small wonder that people often want to use the lowest percentage of any active ingredient. The low-percentage products (10% or less of the active ingredient) are more pleasant to use, but they generally provide good protection for only an hour or so, and partial protection for 2 or 3 hours. Products with more active ingredient (20% or more) often continue to protect against mosquitoes for 6, 8, or even 10 hours. We don't like mosquitoes because of their immediately irritating bites, but this unlovely characteristic does make it easy to tell when it is time to reapply repellent.

Which active ingredient is best against mosquitoes? They all work well; the advantages of one compared to another are somewhat subtle. DEET and DEPA are cheaper and available at higher percentages, but they have an oily feel, a strong odor, and a disturbing tendency to melt plastics, dissolve ink, and ruin furniture varnish. Picaridin lasts a little longer than DEET at the same percentage, has a milder odor, works a little better against malaria mosquitoes, and does not interact much with plastics and varnishes. IR3535 has a history of variable performance, but in the right formulation at 20% it provides eight or more hours of protection against all mosquitoes. The big advantages of IR3535 are that it has little odor and a non-oily feel, and it is extremely safe to use. As we mentioned in chapter 8 on natural repellents, PMD is also a great repellent against mosquitoes. The botanical mixtures known as *Quwenling* or the misnamed "oil of lemon eucalyptus" may be even more effective thanks to the natural addition of other active ingredients. These products appear to be somewhat more effective against malaria mosquitoes. The main disadvantages are the strong odor of the botanical extracts (though pure PMD has little odor) and the niggling concern with skin sensitization and eye irritation.

You can apply repellent to the surface of your clothing when mosquitoes are so numerous that they bite through places where the cloth lays flat across the skin. This treatment is likely to put a lot more product between you and the pest because you can use a heavier application of the repellent and because none of the active ingredient will be absorbed into your body. Spraying outer clothing has been tested most thoroughly with DEET, which lasts several days on cloth. An old technique that soaked a net jacket in 75% DEET actually produced some spatial repellency against mosquito bites elsewhere on the body. Unfortunately, it takes a lot of DEET to achieve this spatial effectiveness. Generally, you will not get much protection of exposed skin on your face, arms, or legs when you spray the shoulders of a shirt or the thighs and knees of trousers, where mosquitoes commonly penetrate.

To repel the small biting flies, you will also have to treat every part of your skin that is exposed. At least you won't need to put repellent on your clothing because none of these small insects can get through. Luckily, the most dangerous of the small flies, the sand flies, are very sensitive to repellents. If you are in sand fly country, you will want to put on repellent before you start getting bites. Sand flies often bite sleeping people who are unaware of the problem until they wake up with dozens of itchy bites, potentially infected with leishmania parasites. Black flies usually bite during the day, and we notice the problem right away. Their irritating habit of crawling about before biting, especially just at the edge of a sleeve or on the scalp, can make them nearly unbearable. All of the repellents work well to stop black fly bites, but they are less efficient at keeping them off your skin. When black flies are a problem, lower-percentage products may limit the number of welts, but they don't seem to stop the flies from crawling about. Higher-percentage products, especially DEET, applied liberally can get you some relief from the crawling, as well as the biting, of black flies. Biting midges did not read the same book about repellents as the sand flies and black flies. Biting midges can be a real plague, some biting during the day, others in the evening or at night. DEET, DEPA, picaridin, IR3535, and PMD all claim effectiveness, but some field trials have shown disappointing results. There is no question that the application of any of these repellents will make a big difference in your comfort level, but don't expect the same duration of protection. Picaridin provided relatively good protection against a particularly bothersome biting midge in Scotland, and DEET provides at least a couple of hours of protection against day-biting black gnats, but there is still a lot to learn about using repellents against biting midges.

The effects of repellents on the larger biting flies are not well studied. Among the horse flies, tsetse flies, and stable flies, the latter have been tested most thoroughly. The results have been mixed, but it seems clear that all of the major active ingredients repel bites for at least an hour or two. The addition of MGK 264 and MGK 326 to DEET extends the repellent's effectiveness considerably. Application to the ankles and legs is usually adequate against stable flies because they tend to bite low. We know much less about protection from horse flies, deer flies, and tsetse flies. Only DEET makes clear claims, but all the ingredients we've discussed probably give an hour or two of relief. Experience with DEET shows that it is worth using against horse flies, but you will end up reapplying often and still getting a few nips from these large biting insects.

Use of repellents for protection from fleas is not a routine thing. Few people would tolerate a flea-infested home for very long without taking measures to stop them at their source (usually pets or rodent infestations). Outdoors, only the dedicated mammalogist or spelunker is likely to run across bothersome populations of fleas. DEET, DEPA, picaridin, and IR3535 make claims against fleas, and they are all probably effective. Good formulations should last at least a couple of hours and maybe longer. Most fleas hop onto the feet or lower legs. The application of repellent to your trousers, ankles, and shoes should help a lot if you need relief in your home while you take care of the problem or if you are one of those dedicated specialists who run across fleas in the field.

WHO IS USING THE PRODUCT?

Although we do not like to admit it, cost is a big factor in choosing a repellent. If your budget is particularly challenged, you may find yourself buying whatever comes in the biggest bottle for the lowest cost. We have discussed in some detail the relative effectiveness of the major active ingredients. They all provide some protection against many of the biting pests. A decision based on cost is not necessarily a bad one; just be careful to consider your requirements in relation to the product. For example, if you need something to provide an hour of protection while you cook the burgers on the barbecue, the big bottle with 10% active ingredient may be good value. On the other hand, if you want something to protect you on a two-week sojourn in the woods where you need protection for many hours per day, a more expensive product that claims six to eight hours of repellency may actually cost you less by the end of the trip. Generalizations about cost are difficult because prices vary according to marketing strategies, locale, and season. You can certainly watch for bargains at the end of the mosquito season, but most of the time DEET and DEPA products will cost the least per hour of protection. A new product containing PMD and lemongrass oil is being designed specifically for those who live in regions that experience both poverty and a high incidence of bug-borne pathogens. Remarkably, this product is projected to be almost 10 times less expensive per hour of protection than the next cheapest product (see Box 9.2 and Box 14.1).

Box 9.2 The Cost of Repellent Products

The pricing of consumer products is so variable that it is hard to be sure of consistent trends. Mark-ups at retail outlets when people badly need repellents and mark-downs at the end of the bug season lead to a cost volatility that rivals the stock market. When you only have so much money in your pocket, buying enough volume for your entire family is the main consideration. At other times, price is no object and you want the most convenient package, the most effective bite prevention, and the best cosmetic properties. In order to avoid false economies, consider exactly how you intend to use the product. If you don't care how you smell and you only need protection for an hour or so while you do chores, a cheap, low-percentage DEET product may be just the thing. That same product may not be a bargain if you need protection all day long and you would require quarts of the repellent to get protection. Although even this rule has many exceptions, you will generally pay more for extra features like sunscreen, great skin-feel, and a pleasant scent.

Table 9.2.1 lists a range of products to give you an idea of prices as of 2007. We assume that a person needs about 5 grams (the weight of a nickel) per application and that label claims of protection times are accurate. You can see that a long-duration product can be a bargain on a per-hour basis, but not if you only need protection for a short time.

Table 9.2.1 The Cost of Some Repellent Products

Product Description	Active Ingredient	Hours of Protection	Cost per Application*	Cost per Hour of Protection*
aerosol controlled-release	25% DEET	6	14	2.2
liquid	20% DEET	12	29	2.4
LIPODEET controlled-release liquid	30% DEET	12	46	3.8
polymer aerosol	25% DEET	6	17	2.8
pump spray	25% DEET + MGK 264 + 326	8	31	3.9
polymer cream	35% DEET	12	83	6.9
cornstarch aerosol	15% DEET	4	31	7.7
towelette	25% DEET	6	52	8.6
liquid	100% DEET	8	78	9.8
sunscreen combination	10% DEET	2	24	12
towelette	5.6% DEET	2	27	13
pump spray	10% picaridin	7	23	3.2
pump spray	5% picaridin	4	18	4.4
aerosol	20% picaridin	8	55	6.9
aerosol	20% IR3535	8	28	3.4
sunscreen combination	20% IR3535	8	35	4.3
liquid	40% *Quwenling* (26% PMD)	6	33	5.5
liquid	16% PMD, 5% lemongrass oil	7	2.6	0.37

*Based on 2007 survey of actual retail prices expressed as U.S. cents.

Most people are quite reasonably concerned about putting repellent on their children. A young child suffers terribly from mosquitoes and other biting pests because the immune system has not yet had time to adapt to the bites. A one-year-old exposed to moderate biting pressure from the common house mosquito will have alarming red welts all over her body, causing a parent to explore desperate measures the very next night. Children old enough to express themselves are less stoic than adults about the irritation of bites and less able to see the pleasures of the outdoors outweighing the pain of mosquitoes, black flies, midges, or other biting pests. What is more, we tend to dress children in thinner, less protective clothing. Since children suffer badly from bites, the temptation is to use a great deal of repellent on them. The problem is that, being smaller, children have more skin in relation to their volume than do adults. This means that what seems like the same rate of application as for an adult is actually much greater in relation to the size of the child. Repellent labels give various directions about the application of repellents on children, warning the parent to supervise use and limit the number of applications per day. Warnings have been given and withdrawn that limited the recommended percentage of DEET to 10% or less for children. The warning was withdrawn in the United States because there was no indication that higher percentages were actually harmful. Although all of the active ingredients are safe when used according to registered label directions, IR3535 certainly has the best safety record and toxicological profile. We would also recommend thinking about how to minimize the amount of repellent you need for your children. In homes with children, it is all the more important to apply the measures described in chapters 4–6. While camping or doing other things outdoors, consider dressing the kids in clothes that protect them better from bites (chapter 7) and make maximum use of treatments to the shoes and outer clothing, as described in this chapter.

Pregnant and lactating women may feel particularly sensitive about applying chemicals that might be absorbed. None of the major repellent active ingredients include official warnings about use by pregnant or lactating women, but common sense would suggest that minimizing chemical contact is a good idea. This does not mean that these women should suffer bites in order to avoid using repellents; it does mean that they should take advantage of other means of personal protection to the maximum extent. DEET is absorbed a lot from many formulations, but it is quickly metabolized and excreted. Other active ingredients are not absorbed as easily, but they have also not been tested as thoroughly (or at least, the tests have not been made public).

WHAT ACTIVITY ARE YOU DOING?

It would be nice if each repellent product worked exactly as it claimed on the label regardless of what a person was doing. Unfortunately, the reality is that the activity and conditions of the person applying the repellent can change its performance dramatically. Most repellents are used during outdoor activity, but sometimes it is convenient to use them indoors for temporary relief from fleas, biting mites, or indoor mosquitoes. If the weather is not too hot and you are not sweating much, repellents will often exceed label claims under indoor conditions, especially if the number of biting pests is not too great.

Outdoors, repellents often last less time than the labels claim. One exception is when the number of biting pests is low and there are alternate hosts (for example, friends who did not use repellent). Under these conditions, if the weather is not too hot and you are not performing a vigorous activity, the repellent might actually last longer outdoors than the label claims. Otherwise, the enemies of repellent longevity will come into play and you are likely to have protection for much less time than you might wish. Repellents are lost from the skin through wetting and through abrasion. Rain and sweat wash off the active ingredient, greatly reducing the duration of its effectiveness. Some formulations resist loss from wetting, but none are completely waterproof. Abrasion occurs whenever anything touches the surface of your skin that has been treated with repellent. As a result, vigorous activity is likely to reduce the protection time. The activities of combat soldiers are probably the ultimate challenge to the duration of repellent effectiveness. While they are constantly touching tools and vehicles, working hard and sweating constantly, crawling through bushes, and maintaining activity in the rain, the very best repellent product known (a polymer formulation of 33% DEET) may provide protection for less than a half hour. It is small wonder that a well-equipped soldier needs permethrin-treated protective clothing, trousers tucked into boots, long sleeves, and repellent for exposed skin. The avid civilian outdoors person should pay careful attention to the formulation-related claims of the products, selecting those that say they will release the active ingredient at a constant rate and will resist moisture. Those products are likely to be more expensive, but the result could be much more satisfactory. Again, all of the active ingredients are generally effective, but picaridin appears to have a slight edge in terms of duration. It makes sense that if duration of repellency is a challenge because of vigorous outdoor activity, the active ingredient that lasts the longest on its own is the right choice.

WHAT IS THE THREAT FROM THE BITES?

Getting some relief from hordes of biting midges or persistent mosquitoes is one thing, but avoiding potentially serious diseases caused by arthropod-transmitted pathogens is quite another. The motivation for avoiding bites should be much greater when there is a significant threat of disease. We have already reviewed strategies for using repellents against each group of biting pest, but it might be good to put these measures in the context of some of the most common diseases associated with bugs.

The ultimate bug-borne infection is malaria. Worldwide, it is one of the biggest killers. Most of the malaria mosquitoes are sensitive to repellents, though all of them seem to be at least a little more tolerant than other mosquitoes. If you are in an area like Korea, Mexico, or Turkey, where the kind of malaria is less serious (though if you get it, you will not think so) and the percentage of infected mosquitoes is low, you might use a repellent much as you would for nuisance pests. If you wait to reapply until you notice bites, your chances of getting infected from those few bites are pretty low. On the other hand, if you are in tropical Africa, Cambodia, or parts of the Amazon, the malaria there is often deadly and a high percentage of mosquitoes are infected. Sometimes, you might find yourself in a highly malarious area without the drugs used to prevent infection. Your reasons for not taking the drugs might vary, and some would argue that there is never a legitimate reason for avoiding the drugs in highly malarious areas, but there can be reasons, such as individual sensitivity to side effects, very brief stays in the region, and concerns over pregnancy (a particularly difficult situation, since the malaria infection itself is dangerous to a pregnancy). Under these conditions, you should apply repellent regularly (at least every four hours for the best products), starting at sunset and ending at dawn, in order to prevent every single bite. If you are in a hotel with a bed net, you should follow the recommendations in chapters 5 and 6, but also use repellent as an extra protection for parts of your body that might contact the net while you sleep. Be especially careful if you are eating outdoors at night. Distracted by the activity, you may not even notice the mosquitoes busily infecting you with malaria. Presented with this sort of situation, it is important to have the right repellent with you. Be sure you have something you can spray on easily if you find yourself outdoors at night unexpectedly. A repellent with low odor (that is, not DEET, DEPA, or oil of lemon eucalyptus) might be important if the outdoor occasion is more formal.

Dengue is another common infection in the tropics. The virus is transmitted by day-biting mosquitoes. Be careful of statements that they bite more often in the morning or in the afternoon. The fact is that the time of day when they are most active depends on temperature, so that on relatively cool days they will be most abundant in the late afternoon, whereas on hot days they avoid biting at midday. Be careful to spray your trousers, lower legs, ankles, socks (they love black socks), and exposed feet because these mosquitoes like to bite low. You are particularly vulnerable when you are eating in an outdoor or semi-enclosed restaurant, though these mosquitoes readily go indoors to bite. Fortunately, the infection rate of the mosquitoes is usually low. Your risk becomes a game of chance: the more bites you receive, the greater your chance of becoming infected with dengue virus.

Tick-borne pathogens cause some scary illnesses, like Lyme disease, Rocky Mountain spotted fever, and Congo-Crimean hemorrhagic fever. Individual populations of ticks often have a high rate of infection, increasing the need to prevent bites. The very small, larval ticks are not usually a source of infection, but the next stage can be important in transmission. This next stage, the nymph, is not as tiny as the larva, but they can be hard to see, and therefore hard to remove, before the pathogen makes its way from the tick to your body. Adult ticks are conspicuous. If you are careful to examine yourself a few times a day, you will be able to pull off adult ticks before they are a danger. Those mechanical measures are important because repellents are not particularly good at preventing tick bites. Protective clothing, permethrin-treated clothing, and the application of repellents as a barrier on the tops of shoes and on the lower legs should prevent most ticks from attaching.

There is more to a repellent product than its active ingredient. The way the chemical is mixed and packaged will change how you use it and affect which product is the best one for you. The next chapter presents some of the things you might consider before you invest in that bottle of bug juice.

FORMULATION

The Choice Is Up to You

Time and again, we have mentioned that the active ingredient in a repellent is only part of what goes into the quality of the product. Other components of the repellent can have a big effect on performance, skin-feel, and absorption. Even packaging can make a difference in how people use a product.

PACKAGING

Let's start with packaging because your first choice may be more about how you want to use the product than what is inside the repellent.

Aerosols in pressurized cans are popular for some obvious and not-so-obvious reasons. First, they are easy to use in that you only have to press on a valve to create a fine mist. Second, and possibly most important, the combination of a fine spray and a push-button gives you excellent control over the location and amount of application. Third, sprays are the most practical way to treat clothing in order to achieve an even distribution of the product. Finally, you can apply an aerosol from a pressurized can without getting your hands covered in the repellent. It's easy to spritz a little more aerosol to stop the bites.

Aerosols have some disadvantages, too. The can must contain a pressurized propellant as well as the active ingredient in a rigid container, resulting in a more bulky package. That package is subject to additional restriction because it is pressurized. It cannot be packed in carry-on luggage if it is larger than a minimum size, and it sometimes raises security questions even in checked baggage. Leakage can result if the aerosol gets too hot, a circumstance that is all too likely in a car during outdoors activity. Both common sense and the label direct you to avoid spraying the repellent directly onto your face in order to prevent contamination of the mouth, nose, eyes, and ears. Some labels state that the product should only be used outdoors where ventilation minimizes inhalation of the product during application. The most subtle disadvantage of aerosol products is that a person tends to put on much less active ingredient, resulting in less effective performance than expected. An aerosol can is more expensive to make than a simple bottle, but the consumer is unlikely to notice the difference in price for the most popular, bulk-manufactured products.

Pump sprays are almost as convenient as pressurized aerosols. They produce a fine, even mist of the product, but they require the constant depression and release of a button or lever at the top of the container. There are a variety of mechanisms for these sprays and they vary in quality. You might have a good one that actually gives you more control than with a pressurized aerosol, or you might have a bad one that tends to stream and dribble. With nonpressurized pump sprays, you may have problems when you try to apply the formulation to the lower limbs. You will tend to tip the bottle upside down, which results in poor flow or no flow when air enters the outlet. If the spray top is poorly constructed, you may also get some leakage of the repellent. A pump spray does not have to be packaged in a rigid container, and the same weight of product fits into a more compact bottle. The pumping mechanism is not as simple as the valve at the top of a pressurized aerosol, however, making pump sprays a little more expensive. They are a good solution for travelers who want to avoid aerosol cans in their luggage, though leakage from a pump spray is always a worry unless it has a tight cap that fits over the pump mechanism.

Liquid repellents in plastic bottles are another common type of product. The bottle usually has some sort of pinhole opening, either as a simple insert or as a small flip-up valve. The pinhole is a considerable convenience because it gives you much better control over the quantity of material dispensed.

The bottle of liquid is compact and leakproof, but it is hard to use such products for even application to clothing. Liquids have a considerable disadvantage compared to sprays and aerosols because the application of a liquid always involves getting repellent all over your hands.

Tubes are usually used for thick, cream formulations. The hole at the end of the tube may or may not have a flip-up valve, but the thickness of these repellents allows you to dispense exactly how much you want. Like liquids, creams in tubes give you a very compact container—some even getting smaller as you use the product. It's not impossible, but it is difficult to apply creams evenly to clothing.

Towelettes are tightly folded paper napkins soaked in a repellent formulation and packaged in chemically resistant, sealed packets. They are popular because they can be purchased one at a time or handed out to individuals instead of passing around a container of repellent. Application of the repellent is efficient because people are accustomed to using towelettes for hand cleaning. They work well for applications to small areas, but they don't usually have enough material to cover exposed legs, arms, and face.

At one time, repellents in a solid stick or roll-on were popular, but now they are rarely seen in common brands. Both are bulky in rigid containers, but they are easy to transport in luggage or camping gear. Most important, they can be applied exactly where you feel you need them without getting repellent on your hands.

FORMULATION

A repellent is a lot more than its active ingredient. We have discussed packaging, but it's what's inside the package that we actually use. The active ingredient is generally less than 30% of the product, and the remainder of it is usually designated "inert ingredients." Inert they may be, but they are anything but inactive. The other components of a repellent product dilute the active ingredient, distribute it on the skin, control its rate of evaporation and absorption, and influence how the product feels and smells. If the active ingredient is the astronaut who grabs all the attention, the substances making up the formulation are the vessel that got him into space and keeps him alive.

The simplest formulations merely dilute the active ingredient and provide a medium that allows us to spread or spray the material over the skin. Most commonly consisting of primarily alcohol, these formulations do nothing to change the rate of evaporation, and therefore duration, of the active ingredient. In the case of DEET, at least, alcohol actually increases the rate of absorption into the skin. That rate of absorption can be so high that it accounts for a significant decrease in effectiveness as the DEET disappears from the surface of the skin. The old U.S. Army repellent was made of 75% DEET and 25% alcohol. It put a lot of active ingredient on the skin, but it did not last as well as the more recent formulation with only 33% DEET. On the other hand, if you mixed the old repellent with peanut butter, it made a great fuel for heating coffee.

Box 10.1 LIPODEET: A High-Tech Formulation

This chapter emphasizes the importance of formulation on the function of repellent products. Industrial, government, and academic laboratories have worked on this problem for years. In the early twenty-first century, one result of systematic and scientific research on the formulation of repellents is LIPODEET. LIPODEET consists of tiny droplets of DEET surrounded by lipid (fat) in a water-based matrix. This system has been used for years in cosmetics, but it took some doing to develop a preparation of 30% DEET that was stable during storage and cheap enough to manufacture on a practical basis. Early reports on this preparation were published in the scientific literature, and it sounded nothing short of miraculous. Using animal models, researchers saw protection times exceeding 24 hours. Further studies revealed a second advantage: LIPODEET was not significantly absorbed into the bodies of the animals. Working properly, the formulation binds to the outer, nonliving layer of skin. DEET is released slowly from this layer and is eventually sloughed off with the dead skin in a normal bodily process. Excellent duration of repellency and the elimination of absorption into the body are qualities that make a uniquely advantageous DEET product. In addition, the material rubs into the skin like the best quality hand lotion and resists removal by wetting. It will be interesting to see how the public responds to this new high-tech product and whether this process can be adapted to other active ingredients and to aerosol application.

Liquid formulations can get a lot more sophisticated than simple alcohol solutions. Generally, any addition thickens the liquid into something more like a lotion, but that is not a bad thing from a purely functional standpoint. Thicker materials force you to put more volume on your skin, increasing the dosage of active ingredient. Thicker liquids are more difficult to formulate into sprays—either pressurized aerosols or pump sprays. There are limits, but technology has proven equal to the task by inventing new kinds of mechanisms that handle more viscous liquids.

It is hard for the consumer to know what's inside a lotion-like repellent. Most labels do not list the ingredients of the formulation, placing you at the mercy of label claims and direct observation. At the low end of the spectrum, a lotion might consist of the cheapest thickeners and a fragrance to mask the smell of the active ingredient. These products can be a big improvement over simple alcohol solutions, slowing evaporation and the absorption of the active ingredient and making a more pleasant product. Combined with essential oils, these relatively simple formulations can achieve the same duration with a quarter of the active ingredient. Products at the high end of the spectrum integrate some of the most sophisticated material and cosmetic science to produce tiny packets of repellent in a matrix that slowly releases the chemical outward at the same time as it prevents absorption inward. Combined with fragrances specifically selected to neutralize the smell of the active ingredient, these products are among the best formulations available. Naturally, when companies have gone to that kind of research and expense to improve a product, they are not shy about trying to explain their success to the public. Rather than simply stating that the product "works" or "provides hours of protection," they will have labels and Web sites that cite data, explain mechanisms, and generally give you cause to believe they are really on to something.

Any repellent can be made into a cream by adding appropriate thickeners, but who wants it? A cream usually requires a bit of effort to create an even layer on the skin. The hands are left with a layer of the stuff and, if it has an oily base, the layer of repellent feels greasy. Early work to produce long-lasting repellents mixed a series of large molecules known as polymers into cream formulations. Some marvelous formulations were developed that used low percentages of DEET, stopped mosquitoes from biting for over eight hours, and, unfortunately, left a solid, cracking layer of gunk on the skin. Eventually, a serviceable cream was developed that was hard to apply but very effective with only 30%–35% DEET. Other polymer formulations have been marketed, and they definitely decrease absorption and increase duration. The current U.S. military repellent (not-so-euphoniously called EDTIAR—the extended duration topical insect and arthropod repellent) is such a polymer formulation in a tube. Polymer formulations are also available in sprays, though the duration is somewhat reduced. Yet another approach was the use of a cellulose polymer as a gel—which is great if you like to apply something slimy. Polymers tend to resist being washed off by sweat or rain. They are so thick that the dosage applied tends to be greater than that provided by a liquid or lotion, and it leaves a reservoir of product on the skin in case some of the repellent is rubbed off by abrasion. Current recommendations are to wash repellents off the skin once you return indoors, making a waterproof repellent somewhat problematic.

Combination products serve two functions, usually combining a repellent with sunscreen. Manufacturers have latched onto the idea that people who want to enjoy the outdoors free of the dangers of sunburn might also want to be protected from day-biting pests like ticks, black flies, and mosquitoes, such as the Asian tiger. You can get products that combine DEET, IR3535, or other active ingredients with a sunscreen, often in a formulation that is specifically designed to resist being washed off. Some of these products are popular; it makes sense to avoid the effort of putting on two different lotions, not to mention the expense of buying two different products. The problems are that DEET, at least, decreases the effectiveness of some sunscreens, and the duration of each kind of product is different. The best products have dealt with the first problem by testing their combinations carefully for sunscreen protection. They have dealt with the second problem by attempting to match the percentages of the active ingredients of sunscreen and repellent so that they last about the same amount of time. The professional community is conflicted about the use of repellent-sunscreen combinations. On the one hand, major brands receive full registration for their use. On the other hand, the U.S. Centers for Disease Control and Prevention very specifically recommends not using them, mainly because of fear that either the sunscreen or the repellent will last longer than the other, leaving the consumer exposed to either sunburn or bites. As usual, following label directions is the best practice. Unless there is a warning to limit the number of applications, you can always put more on if you start to burn or if the bugs start to bite.

Other combination products include camouflage face paint with 30% DEET and soap with DEET and permethrin. The camouflage face paint is a specialty item if there ever was one, but hunters and the military use the stuff quite often. It wasn't easy getting a product that would hold 30% DEET and still retain the best characteristics of face paint. Literally years of research went into achieving this feat of formulation—a particular example of the effort sometimes required to develop a repellent. The repellent soap is intended as a low-cost, easy-to-use solution to malaria and dengue control. It is one of the few repellents documented to reduce the transmission rate of malaria (in Afghanistan), but it has two disadvantages. First, the intended application involves lathering the soap on your skin without rinsing—not a pleasant thought for those who have experienced the water cutting off in the middle of a shower. Second, the routine application of permethrin directly on the skin is considered hazardous by some toxicologists. You may encounter other combination products that have not enjoyed as much marketing and testing, but it is hard to know how well they will work.

COSMETIC APPEAL

We might think of the skin-feel and odor of repellent products as secondary to their ability to stop bites. However, a repellent doesn't do anyone any good if it stays in the bottle. The ideal product should be pleasant on the skin and have an inoffensive odor. The cosmetic industry knows a lot about making nongreasy hand creams, easy-to-apply foundation creams, and pleasant-smelling or odorless products. The best repellent products take advantage of that technology; the worst repellent products depend on the absolute desperation of their customers to apply anything that will stop the bites. We recommend that you get a product that not only serves your intended purpose, but that also appeals to how you want your skin to feel and how you want to smell. If you are comfortable with a repellent, then you are more likely to use it when you need it—even if it does not last as long as a less pleasant product.

Manufacturers are motivated to produce repellents that are pleasant to apply. Advertising and label claims bombard us with phrases like "pleasant fragrance," "nongreasy," and "leaves the skin feeling fresh." Both the active ingredient and the formulation can contribute to these claims. For example, one popular product uses 5% picaridin and truthfully claims that it will give you a "clean feel." Another product uses new technology in an aerosol to integrate cornstarch and DEET, leaving the skin amazingly smooth and dry compared to alcohol-based spray formulations.

Gene Gerberg, one of the grand old men of repellent development, quipped that, when it comes to repellents, one size does not fit all. He was absolutely right; the choice is up to you.

HOW DO I KNOW WHAT WORKS?

Biting pests are in the business of finding blood. They are taking advantage of a resource that is in great quantity compared to their size, but the path from wherever they are to completion of the blood meal is full of challenges for them. The biting pest has to realize you are there, find you, locate a place on your body that is exposed and suitable for the bite, and find the place in your skin where it can get blood. This complex process, which may very well end in a smashed bug on your arm, is influenced by what the insect sees, feels, smells, and tastes. A combination of signals might say "no blood here," or it might say "a juicy source," or something in between.

Biting pests are not automatons that always do the same thing. The biting preference of an individual insect, tick, or mite at a particular time and place is going to be the product of a number of internal and external factors. The age of the pest, the weather, the time of day, the time of year—all these things make a difference. Most people notice that one person or another in their family seems to get most of the bites when everyone is together. That is no illusion; some people are definitely more attractive than others to biting pests. Your appeal to biting pests will depend on how hot and sweaty you are, the cologne you wear, what you eat, how recently you bathed, and perhaps most important, the genetic inheritance that predetermines the personal chemistry of your blood, your skin, and your breath.

The combination of internal and external factors that influence the biting arthropod and the bitten host determine whether a particular pest bites a particular person. A rough tabulation of these factors creates a complicated picture of what bites whom (Table 11.1). When you think about it, many of the same factors affect the bug and the human. For example, high humidity often favors activity by mosquitoes, but it also tends to leave a person sweatier and more attractive to biting insects. Arthropod repellents may prevent all bites for a while, but the complex interaction of the biting pest's blood lust and the qualities of all the potential hosts within striking distance will determine exactly how long and how effectively the repellent does its work. As a result, protection times stated on a label will sometimes overestimate or underestimate actual performance.

Entomologists do not precisely understand the interactions of these factors and are unable to predict exactly who will get

Table 11.1 Factors Affecting Whether or Not an Individual Biting Pest Sucks the Blood of a Particular Person

Factor Affecting Pest and Potential Host	How the Factor May Affect Appetite or Attraction
age	older pests can be more aggressive; adult people are larger and therefore more attractive
size	larger mosquitoes of the same species can be more effective blood feeders; larger people are usually more attractive
temperature	higher temperatures usually mean more active pests; hot, sweating people are usually more attractive
humidity	some pests bite much more readily in humid conditions; sweating humans are usually more attractive
infection	some viruses cause mosquitoes to bite more often; some diseases cause people to be more attractive to pests
genetics	some pests inherit more aggressive feeding habits; some people are inherently more attractive to pests
location	if the pest is close, it is more likely to bite
physiological state	biting pests become more or less hungry during the egg development cycle; humans vary in attractiveness
time of day	some pests bite at particular times of the day; a sleeping person is very susceptible to bites
activity	sometimes disturbing pests stimulates them to bite; a moving person can be harder to bite or generate more attractive chemicals by sweating
light level	pests vary in what light level they prefer for biting; humans have trouble swatting a pest in the dark
chemical exposure	insecticides and repellents discourage the pest; repellents on the skin limit the biting, but some skin care products are attractive
nutritional status	a pest with a recent blood meal is less likely to bite; various nutritional influences seem to affect biting by some pests
chemical signals	some pests secrete chemicals that attract others to bite; humans secrete many attractive and repellent chemicals from their skin, breath, and blood
species	biting pests have definite preferences for certain animals

bitten by what. Sure, it is clear that some frog-feeding species of mosquitoes never bite humans and that alcohol consumption will increase the attraction of an individual to at least some kinds of biting pests, but no one can say with any certainty that individual A will be bitten and individual B will not. The point is that a repellent that works great for one person under one set of circumstances may not perform as well for another person or under a different set of circumstances.

DURATION

Repellent products usually express how well they work by stating the kinds of pests they repel and the number of hours of protection to be expected. Sometimes, they will make a different claim for different kinds of pests. The expectation is that, if you apply the repellent as instructed, then you will get the stated number of hours of protection. In practice, the number of hours of protection will tend to decrease if you are active, because you will rub more of the product off your skin and sweat will wash it off. If you look down the column of the table of factors that influence the attractiveness of a human, you can imagine that anything that increases attractiveness will make the job of the repellent that much harder. Although it could be caused by diet, physiological status, fragrances, or other temporary influences, we all notice that some people tend to get bitten more than others. Those unfortunately attractive individuals need to reapply the repellent products more often to maintain a protective barrier. On the other hand, people who are located near those susceptible, bite-prone companions will notice a distinct relief from biting pests. They may notice that repellent applications remain effective longer than usual while they are with their less fortunate friends.

The conditions of the biting pest will also make a big difference in the duration of the repellent product. A repellent is likely to last longer during the part of the day when the biting pest least likes to feed, for example. Cool or very hot weather, as well as windy conditions, can discourage bites. If the main problem is a mosquito that prefers to feed on pigs rather than on humans, then the presence of a pig is likely to bring a great deal of relief to nearby people. Misery loves company and so do some biting pests, and increased numbers sometimes seem to overcome repellents when the pest populations are very high. The basis of this feeding frenzy may be the result of chemicals secreted by the pests themselves, but we actually know little about this phenomenon. Of course, if there are enough of the pests around, there are always a few that are unusually tolerant and manage to overcome the repellent as its concentration on the skin decreases.

Governmental organizations have struggled with the evaluation of repellents, both to compare their effectiveness and to provide some sort of assurance to the public that label claims are true. The evaluation task is extremely difficult because of the complexity of the relationship between biting pests and individual people.

Consider the extremes of the spectrum of testing methods to determine the duration of protection from mosquitoes. The control freaks would have a single strain of a single species of mosquito reared identically on the same food in every testing laboratory. At exactly the same age, these mosquitoes would be exposed to identical artificial membranes covering identical mixtures of artificial blood heated to precisely the same temperature. Going to these extremes would increase the likelihood that you could make accurate comparisons of duration between products. But what would these results mean? The comparison would apply to only one species of one age in one physiological state. How representative is that? Worse, the artificial membrane would only model the evaporation of the repellent product from the skin's surface, not its absorption as happens on real skin. At its best, such a test would give you an accurate means of comparing repellent products, but it would tell you little about how much protection time to expect.

People who enjoy outdoor activities would propose field testing. Line up 10 people in a place with natural populations of mosquitoes, treat each person with a different repellent (well, 9 of them: for comparison, the unlucky 10th person gets no treatment), and see how long they are protected. Since each person will inevitably vary in attractiveness, it will be necessary to try each repellent treatment on each individual at least once. You don't have to take off your shoes to recognize that means

10 days of testing. Of course, each day is different—different weather, different mosquitoes, and people in different physiological states. The time of day for peak biting is going to be different for each mosquito species present in the test area, and it is unlikely that any of them will bite continuously with the same appetite during the entire 6–14 hours of testing required. Therefore, you will have to set up staggered treatment times and expose people simultaneously who have had repellent on for different periods. Oh, and did we mention that this must be done either where mosquitoes do not carry disease-causing pathogens or where the disease from the pathogens can be prevented by vaccinations or drugs? The whole task is not impossible and is actually done routinely by academia, industry, government, and testing laboratories. As you can appreciate from this short description of the procedure, "not impossible" is not the same as easy. At the end of this expensive and demanding process, you are only starting to understand the effectiveness of the product for the complete population of customers. It is small wonder that so few field tests have been performed against biting pests other than mosquitoes.

Between the two extremes of testing are a number of compromises that make things easier. A cage of carefully reared mosquitoes can be placed over an arm or leg and the number of bites recorded. The arm or leg can be placed into a cage of mosquitoes to accomplish much the same thing. Field efficacy repellent tests can be conducted for just a few minutes each hour to get some idea of how long the repellent product protects. We could go on and on describing the variety of ingenious systems that entomologists have developed to test the duration of repellent products, but every one of them is a compromise between practicality and getting an answer.

Of all the biting pests other than mosquitoes, tests against ticks are the most difficult. For one thing, it is difficult to be sure that ticks are not infected with a pathogen, particularly in the field. For another, the intention of a tick to bite is not always easy to determine. Finally, the most practical method for preventing tick bites is not so much to stop them from biting as it is to stop them from crawling on human skin in the first place. For all these reasons, repellent testing against ticks usually eliminates the possibility of the volunteer getting bitten. The best tests are those that look at whether or not the repellent is forming an effective barrier against tick movement. This is done by applying a band of the product around a finger or leg, introducing the finger or leg into a container of ticks for a short period, then determining how many made it across the treated band of flesh. The same kind of test can be used for treated cloth, which is actually a much better way to prevent tick bites.

We have discussed the duration of protection without really defining protection. Governmental regulators tend to treat protection as 100% protection; in other words, how long does it take until the first bite occurs? Often called *complete protection time* (CPT) or "time to first bite," this is a demanding standard for the repellent products. If you have a cage full of 100 mosquitoes, it will be a single unusual mosquito that will determine the CPT. This method is not only somewhat unfair to the assessment of the repellent product, it is also statistically disastrous. The single mosquito represents the extreme of its population, and any estimates of accuracy will be nearly meaningless. Nonetheless, the public intuitively understands CPT, and public health officials understand that only the CPT will eliminate the risk of potential infection (see chapter 3). The disadvantage is that those estimates of duration are not accurate, not even for the purposes of comparison.

PERCENTAGE REPELLENCY

Although the usual comparison of effectiveness revolves around duration, there is another way to compare. Percentage repellency tells you what proportion of the biting population does not bite throughout the repellent application. At 1 hour after application, you might prevent all bites from mosquitoes in your area. At 2 hours, a mediocre formulation might be dissipating to the point that it provides only 89% protection (11 of 100 hungry mosquitoes bite). At 8 hours, the best formulation still prevents all bites; however, at 10 hours, even the best one might be protecting you from only 90% of them.

Percentage repellency is harder to measure than complete protection time because you have to know how many pests would have bitten at that time in the absence of the repellent. The usual method is to have an untreated volunteer count bites or landings compared to a different volunteer who applied the repellent. This design is prone to the inaccuracies caused by variations in the attractiveness of volunteers. If it is a field study, you also have to ask yourself if the two collectors are far enough apart so that hungry mosquitoes deterred from the treated person don't find their way to the untreated person. If the distance is far enough, then perhaps the number of blood-seeking mosquitoes is different in the two locations. One way around this problem is to have the same person serve as his own untreated control by treating one arm and leaving the other untreated. It's not a bad method, but it has its own inaccuracies, especially since volatile repellents will cause some deterrence around the untreated arm. In the laboratory, the use of one treated and one untreated arm is likely to be a very good method.

One big advantage of percentage repellency measurements is that they come much closer to what a consumer actually wants to know. Most people can tolerate a few bites (one study shows that about three bites per hour can be a threshold of annoyance for many people). They appreciate protection even when it dips below 100%. Therefore, a repellent product that gives 100% protection for four hours and 90% protection for an additional two hours might be viewed as effective for six hours by someone who is only concerned with the discomfort of the bites. Of course, if disease prevention is the goal, you can only hope that the 10% of the mosquitoes biting during the last two hours are not infective.

The other big advantage of percentage repellency measurements is that they produce statistically accurate results if the studies are designed properly. Therefore, a repellent product that offers 95% protection at six hours may actually be better than one that offers only 80% protection. In contrast, the difference between six hours of CPT compared to five and a half hours is likely to be meaningless.

IMMEDIATE OR INHERENT REPELLENCY

The duration of protection is the product of the actual repellency of the active ingredient and how long that active ingredient is presented to pests from the skin. Loss occurs primarily from evaporation and from absorption. Both evaporation and absorption are likely properties of repellent active ingredients because they usually must influence the biting pest in the gaseous phase and because they tend to be substances that penetrate tissues. In the last chapter, we discussed how the addition of other substances to a repellent formulation can affect evaporation and absorption, greatly extending or limiting the duration of its effectiveness.

Measurements of the actual repellency of compounds, independent of their rates of evaporation and absorption, are seldom reported in commercial claims about active ingredients. Occasionally, you will see a statement like "works as well as DEET," which may be based on the repellency of the compound independent of how long the material lasts on the skin. It is possible to make quite accurate measurements of the surface concentration of a given chemical that is required to prevent feeding by 50% of a population of biting pests. That figure—the concentration of the active ingredient required to repel 50% of the population of the particular pest—is often valuable for comparison of new active ingredients because it is the most statistically accurate measurement. When we know the duration of a chemical on skin, the immediate repellency can be an accurate and quick means of determining the comparative effectiveness against a variety of pests. For example, we know a lot about how long DEET lasts on the skin in various formulations. We also know that it is effective against most mosquito species in terms of duration. If the measurement of immediate repellency against a previously untested pest shows that a small amount is required to repel 50% of the population, then we can conclude that DEET will be a useful active ingredient against the new pest.

CLAIMS ON THE LABEL

The most basic claims of repellent products are that they protect against certain pests and that they continue to stop bites for a certain length of time. What about other claims? With the exception of combination sunscreen products, the labels usually give little information backing up the qualities of the product inside. Occasionally, a reason is given for the claim, such as the use of a new active ingredient that results in a less oily feel on the skin. More often, the label presents the consumer with a statement of quality to be taken on faith, without data. Eventually, you will discover for yourself whether the claim is accurate, but you need to make a choice in the store and it would be nice to avoid buying multiple products in order to try them out.

One of the statements you may see is that the formulation is "strong" or the "greatest strength available." The claim may be perfectly true, especially when it concerns active ingredients that can be formulated at very high percentages. It is confusing, however, because some of the newer active ingredients are limited to relatively low percentages in some countries (for example, a maximum of 20% picaridin or IR3535 in the United States), whereas others are often formulated at concentrations ranging from less than 10% to nearly 100% (DEET and oil of citronella). Therefore, 20% picaridin is a "strong" formulation that will probably work much better than 10% picaridin, and the word *strong* is correctly associated with improved effectiveness. On the other hand, you don't get much extra benefit from applications of DEET above 50% (and you have a considerably greater risk of irritation). Therefore, in the case of strong formulations of DEET, the word refers only to concentration and has less relevance to its performance as a repellent.

Another qualitative claim is that the product is the "best" or "better than" other products. This superlative is usually applied to duration. If the label just says it is the best repellent with no explanation of why, then the potential buyer can either believe the statement or not. Often, however, the label goes on to mention that this product performed better than others in tests by some third party. This claim is much more convincing, especially if the third party appears to be an unbiased institution. The scientists can argue over the conduct of each set of studies, but at least the comparison is based on data. The real problem is that, for the reasons discussed above, the results of good-faith evaluations can vary with the selection of human volunteers, species of biting pest, conditions of the test, etc. There is at least the possibility that a company will choose the evaluation that most favors its own product. If the repellent did not perform as well as others, the label might only say as "tested by" or "used by" some major institution. The implication is that the institution used the product out of preference, though it is likely that the product was rejected following trials or that the manufacturer simply sent samples.

Controlled release, slow release, timed release: these phrases all suggest that careful work has gone into the formulation to optimize protection from the active ingredient. Such repellent products may cost more based on the claim. The quality repellents are probably well worth the added expense if you want the maximum protection with the minimum active ingredient, but how can you determine which are the quality repellent products? One way is to look at the percentage of the active ingredient and compare it to the label claim for hours of protection. Assuming that the duration claim is either honestly presented or governmentally regulated, a repellent product with 35% or less of its active ingredient that offers eight hours of protection can legitimately state that it has found a long-lasting formulation. A few words on the label may be another hint that there is some science behind the formulation. The words "liposome" or "polymer" indicate that some pretty sophisticated cosmetic science may have gone into the repellent product. Even the advertisement that goes something like "years of research have gone into this formulation" may be a genuine indication that the manufacturer actually did some testing and optimization. All that said, we have to admit that some standard lotion-like bases that required little research for formulation can result in a much longer lasting repellent product than simple alcohol solutions of active ingredients.

Some repellent products' labels, especially sunscreen combinations, state that they resist loss from sweat or wetting. Standardized tests exist for evaluating the water resistance of cosmetic products and sunscreens, but these data rarely find their way into the open literature on repellents. Some of the sunscreen combinations apply the same technology for waterproof sunscreens to their repellent-sunscreen products, but it is not clear whether this strategy works. In fact, there is little indication that any of the formulations actually resist wetting under practical conditions. An alcohol solution of DEET lasts no more than 15 minutes in a heavy rain, in contrast to a polymer formulation, which continues to protect a person for at least an hour. The same polymer formulation may provide protection for only a few minutes if it is exposed to rain *and* rubbing on vegetation (as experienced by some soldiers on patrol). Data on liposome formulations applied to animals suggest that this type of formulation does a very good job of protecting the active ingredient from water, but we have not seen information based on people actually using the repellent product. The bottom line for the consumer is that, when it comes to the water resistance of repellents, you had better experience it yourself.

USE EXPERIENCE

It is expensive to try a lot of different repellent products, but it also makes no sense to buy the wrong repellent and punish yourself by not using it. You can be your own best test subject, especially if you want to use the repellent for an activity you perform often. After considering how you want to use the repellent product, what you want to protect yourself from, and whether or not there is a risk of disease, you should be able to make some choices based on the guidance in this and the previous chapters. Try the repellent product and ask yourself these questions:

1. Did the repellent stop all or nearly all of the bites?
2. Did it seem like the biting pests didn't even get close?
3. Did the repellent provide adequate protection for a long enough time?
4. Was the repellent product comfortable to put on?
5. Was the odor from the repellent tolerable?
6. Did you use the repellent product consistently, whenever it was needed?

If the answer to all of these questions is yes, then you have a good repellent for that activity. Let's take a look at what to consider if the answer is no to any of these questions.

It is difficult to determine the number of bites that exceeds your personal comfort threshold or that creates a significant threat of infectious disease. In a situation where even a single bite has a good chance of infecting you with a disease-causing pathogen, you really do want 100% protection. Even if disease is not a problem, a single bite might be a nuisance if you are at a formal occasion or if you are trying to accomplish a delicate task. On the other hand, people on a canoe trip through beautiful but bug-infested waterways may gladly tolerate 10 bites per hour because they accept that level of discomfort as part of the experience. In general, if you are getting too many bites after applying a repellent, you can try waiting 10 or 20 minutes between the application and exposure to the bugs. Some formulations, especially of IR3535, need time to release the active ingredient. Another possibility is that you are not applying enough repellent, so read the label carefully and follow directions. Finally, you can try switching to a different active ingredient, to a repellent product with a higher concentration of the ingredient you are using, or to one with a mixture of active ingredients.

Some active ingredients are more volatile than others, preventing bites and also keeping the insects away from your skin. It won't make a difference if the biting pests in your area are not the kinds that crawl maddeningly on the skin. On the other hand, some pests are almost as annoying when they are walking on your skin as when they are biting. Black flies are terrible about this habit, as are some species of mosquitoes. A more volatile repellent active ingredient like DEET could be the best choice.

Before you start blaming a repellent for not lasting long enough, make sure that you apply an adequate amount on the skin in the first place. Another possibility is that you are assuming that the repellent owes you protection for the maximum amount of time listed on the label. For all of the reasons we have already discussed, you may not get such long protection. It may be necessary to reapply more often (though be careful not to exceed the recommended number of applications). Special formulations like liposomes and polymers are likely to extend duration.

A greasy, thick formulation is hard to apply evenly, leaves the hands a mess, and generally discourages a person from using it. In contrast, a cooling spray or a smooth lotion practically begs you to use one of them. It is worth the extra money to get a repellent that you as an individual think is easy to use and feels good on the skin. Otherwise, there will probably be times when you think the repellent is worse than the bites.

People also perceive odors very differently. Some think that DEET has a clean chemical smell and they don't mind using it. Others think DEET smells awful and they associate it with uncomfortable conditions and bad times. You can usually get some idea of odor in the store, even without opening the container. Some repellent products are scientifically formulated to cancel the odor of the active ingredient with a carefully selected fragrance. In general, IR3535 and pure PMD have less odor than picaridin and much less odor than DEET.

Probably the best evaluation of a repellent product is whether you use it consistently. If you find that it is easy to use and that it does a good job of protecting you from the pests that bother you, then you will probably apply it whenever you think you might be bitten. Such a product is ideal for your particular situation.

HOW DO I KNOW THAT IT'S SAFE?

> All things are poison and nothing is without poison; only the dose permits something not to be poisonous.
>
> —Paracelsus

People worry a lot about chemical exposure. We are exposed in many different ways: the food we eat, the air we breathe, splashes of this or that solvent as we fill our gas tanks or paint our dresser drawers. The fact is that we are made of chemicals, we eat chemicals, we breathe chemicals, and our skin separates us from the chemicals in our environment. Problems occur when we are exposed to too much of the wrong chemical.[1] *Toxicity* could be defined as the negative action of a chemical on our bodies, but the level of toxicity depends very much on how much of the chemical is involved and how it came in contact with us. Some of the most innocuous chemicals can be toxic. You could kill a person by feeding her too much salt, baking soda, or sugar. Carbon dioxide passes in and out of your lungs with every breath, but put your head in a tub filled with dry ice (which is frozen carbon dioxide) vapors, and you will be dead before you know it. The spicy salsa that you enjoy on your tortilla chips would cause plenty of irritation in your eyes or on other sensitive parts of your anatomy.

Perhaps what bothers people the most about chemical exposure is the fear that some unnoticed substance is causing

1. The quotation above, which makes the same point, is attributed to Philippus Theophrastus Aureolus Bombastus von Hohenheim, more commonly known by his self-appointed title, Paracelsus. Paracelsus was an active Swiss man who practiced the contemporary form of scholastic science as he wandered from institution to institution throughout Europe from the time he was 16 until his death in 1541. Before experimental medicine, people like Paracelsus drew their conclusions from pure reason, other writings, and experience. The system had its limitations (he died at the age of 48), but Paracelsus was right when he implied that the dose makes the poison.

a devastating effect. Could Aunt Sally's cancer have been caused by a careless exposure to a pesticide? Is my persistent acne the result of childhood exposure to chlorinated compounds? Did a preservative in a vaccine cause my child to be autistic? There are so many claims and counterclaims, the data are so complex, the statistics so difficult to understand, and the possibilities for harm so endless that it is only natural that people begin to think that any chemical exposure could result in illness, ranging from nervous disorders to infertility to cancer.

ROUTES OF EXPOSURE

The amount of a substance to which you are exposed is only part of what determines whether or not the material is harmful. The other aspect of toxicity is the *route* of exposure, generally categorized as oral, dermal, or inhaled. Each route of exposure has its own characteristics that influence the effective toxicity of the compound. Eating or drinking a potential toxicant exposes the chemical to a series of challenges that break down many of them. That's why you can eat such a wide variety of things but end up with, for the most part, the same simple compounds penetrating your intestines and becoming more of you. On the other hand, once in the stomach, a resistant chemical is going to be inside for a long time and will have many opportunities for absorption. The skin itself often stops a large volume of chemicals from even entering your body; they either drip off or get washed off when you take a shower. Some chemicals can get through the skin, however, and they then go directly into your system almost as though you had injected them. Other chemicals have a severe effect on the skin itself that they would not have on the stomach; common examples include the wide variety of botanical derivatives that cause sensitization or phototoxicity. Inhalation might sound like the most dangerous route of exposure because your lungs have little ability to prevent uptake. Like the intestines, the lungs are built to absorb things. Good absorption is needed for oxygen but dangerous for an inhaled toxicant. The lungs themselves can be a target for many toxicants, causing two conditions we particularly fear: emphysema and lung cancer. Fortunately, potentially toxic inhaled chemicals are often diluted quickly by the air around us.

IMMEDIATE TOXICITY

Repellent active ingredients would not be useful if they were very toxic, but let's take a look at some of the considerations on their use with respect to each route of exposure. Although there have been various attempts to develop oral insect repellents, we currently do not have an effective product. Other than the intentional ingestion of a repellent in a suicide attempt, oral exposure to repellents is usually accidental, as when a child drinks a liquid formulation or when a tiny bit of the product passes the lips during sloppy application. The active ingredients in repellents are definitely harmful if you drink enough of them. Based on tests with rats, drinking about 7 ounces (200 g) of pure DEET would kill half of the adults who weigh 150 pounds (68 kg). The fatal dose of picaridin would be over 11 ounces (320 g), of DEPA over 2 ounces (63 g), of PMD about 6 ounces (164 g). IR3535 has very low toxicity; even 12 ounces per animal (340 g) failed to kill half of a rat population. In comparison, the corresponding figures for the most toxic commercial pesticides can be as low as a hundredth of an ounce (about a tenth of a gram). Aspirin is much more toxic than most insect repellent active ingredients, requiring less than an ounce (17 g) to kill most adults.

After being applied to the skin, repellents usually evaporate into the gaseous phase before they affect biting pests. Volatile action is an advantage because it keeps the pests at a greater distance from the skin, minimizing annoyance from insects that land before being repelled. The functional disadvantage of volatility is that the repellent may not last as long. There is a toxicological hazard to volatility as well. The more a compound tends to evaporate, the greater the inhalation exposure. Regulatory agencies are keenly aware of this form of exposure, though the public may be less concerned. Realistically, a person using DEET outdoors would breathe in very little of the material. To give you some idea, consider a 5 gram application of 30% DEET that continues to repel insects for six hours. At least 20% of the chemical will be absorbed and metabolized, leaving a total of 1.2 grams of active ingredient that evaporates during six hours. If we assume that a personal atmosphere is 2 cubic meters and that the air around us changes every minute, then the DEET is diluted into 720 cubic meters of air. The average concentration breathed in by the person who applied the repellent would be less than 2 milligrams per cubic meter, 2 parts per billion, or two ten-millionths of a percent—that is, not much. It takes 6 grams of DEET per cubic meter to kill half the rats breathing such a concentration for four hours, or approximately 3,500 times as much as we experience following a thorough application of a product. The comparable inhalation toxicity is 4.4 grams per cubic meter for picaridin and 1.5 grams for DEPA. The bottom line is that, as long as you are using a repellent outdoors, your risk from inhalation toxicity is truly miniscule.

Repellents are, of course, designed to be applied directly on the skin. Toxicological tests really concentrate on this route of exposure because potentially hundreds of millions of people could make billions of applications of an active ingredient. The kind of toxicity we've been discussing so far—the kind where the subject suddenly drops dead—requires so much material applied to the skin (dermal exposure) that it would be difficult to stick it on there. For DEET, it would take almost a half pound (204 g), for picaridin and PMD more than four-fifths of a pound (340 g), and for DEPA over a half pound (270 g). The amount of IR3535 to kill a person would be impossible to load onto the skin even by covering the whole body with it.

But death is one thing and irritation is another. All of the active ingredients, and particularly PMD, irritate the eyes. The sensitive skin lining our orifices is also subject to irritation. Most people are careful to avoid getting repellent on these delicate areas, but sweat, rain, or hands can inadvertently carry some of the active ingredient to even the most remote regions of the body. The result is usually no more than stinging and reddening, but still is very much to be avoided. Repellents can irritate regular skin sometimes to the point of blistering. This kind of skin problem occurs in some individuals on areas that are moist and covered, as happens on the inside of the elbow or behind the knee.

DELAYED TOXICITY AND TOXICITY FROM REPEATED APPLICATIONS

Cancer! Nerve damage! Birth defects! Infertility! These delayed, serious effects of chemicals are greatly feared by most people. The problems are terrible in themselves, but they are made worse by the fear factor stemming from a helpless sense that no amount of care can completely eliminate the risk. There is also the suspicion that normal safe exposure repeated over a long period will have a cumulative detrimental effect.

Manufacturers design repellent products to avoid any harm that might result from repeated use. After all, they are hoping that you will use the product frequently up to the maximum amount allowed on the label. Losing customers to any kind of harm caused by the product cannot be in their business plans. As you can imagine, the tests to assure this kind of safety are complicated, lengthy, and expensive. For repellents, the tests stress the dermal route of exposure. They consider the pattern of use by most people to attempt to determine how large a dose is safe. Many of the restrictions based on age, frequency of application, and percentage of active ingredient come from these considerations.

We introduced the short-term tests above. These consist of a single dose to an animal followed by observation of the results, whether death, behavioral aberration, or tissue damage detected at necropsy. The longer-term tests either dose the animal repeatedly or attempt to determine the presence of a physiological mechanism that could cause long-term damage.

The simplest tests administer the chemical repeatedly to a group of animals that are observed for adverse effects. The effects might include skin sensitization (development of an allergic response), liver damage, kidney damage, weight loss, etc. The "no observable effect level," or NOEL, and the "no observed adverse effect level," or NOAEL, are similar ways to express the results. Both measurements are estimates of the maximum exposure that does not cause harm. The NOEL or the NOAEL, reduced by an appropriate safety margin, can be used to develop a dosage for humans that is considered safe.

A long series of tests examines the chemical's potential to harm the genetic mechanisms, through damage to either

Table 12.1 Standard Toxicological Tests as Described by the European Union's European Chemical Bureau

acute oral toxicity: fixed-dose procedure
acute toxicity (inhalation)
acute toxicity (dermal)
acute toxicity: dermal irritation/corrosion
acute toxicity: eye irritation/corrosion
skin sensitization
repeated-dose (28 days) toxicity (oral)
repeated-dose (28 days) toxicity (inhalation)
repeated-dose (28 days) toxicity (dermal)
mutagenicity: in vitro mammalian chromosome aberration test
mutagenicity: in vivo mammalian bone-marrow chromosome aberration test
mutagenicity: mammalian erythrocyte micronucleus test
mutagenicity: reverse mutation test using bacteria
gene mutation: *Saccharomyces cerevisae*
mitotic recombination: *Saccharomyces cerevisae*
mutagenicity: in vitro mammalian cell gene mutation test
DNA damage and repair: unscheduled DNA synthesis: mammalian cells in vitro
sister chromatid exchange assay in vitro
sex-linked recessive lethal test in *Drosophila melanogaster*
in vitro mammalian cell transformation test
rodent dominant lethal test
mammalian spermatogonial chromosome aberration test
mouse spot test
mouse heritable translocation
subchronic oral toxicity test: repeated-dose 90-day toxicity study in rodents
subchronic oral toxicity test: repeated-dose 90-day toxicity study in nonrodents
subchronic dermal toxicity test: 90-day repeated dermal dose study using rodent species
subchronic inhalation toxicity test: 90-day repeated inhalation dose study using rodent species
chronic toxicity test
teratogenicity test: rodent and nonrodent
carcinogenicity test
combined chronic toxicity/carcinogenicity test
one-generation reproduction toxicity test
two-generation reproduction toxicity test
toxicokinetics
delayed neurotoxicity of organophosphorus substances following acute exposure
delayed neurotoxicity of organophosphorus substances: 28-day repeated-dose study
unscheduled DNA synthesis (UDS) test with mammalian liver cells in vivo
skin corrosion (in vitro)
phototoxicity: in vitro 3T3 NRU phototoxicity test
skin sensitization: local lymph node assay
neurotoxicity study in rodents

DNA or chromosomes. Many of these tests take place in cultured cells or bacteria, avoiding the use of whole animals. The creation of new mutations, rearrangement of chromosomes, changes in DNA production, and other aberrations are assumed to imply that the chemical could cause similar harm in humans and animals. The genetic damage would presumably lead to cancer or to reproductive problems.

Whole animals, usually laboratory mice or rats, are used for the evaluation of potential cancer-causing properties (carcinogenicity). Veterinary pathologists minutely examine animals exposed to a chemical for up to two years, looking for any sign of cancer in excess of that which would normally occur. Similar studies determine whether or not the chemical causes birth defects, fetal damage, premature birth, infertility (in males or females), or other signs of reproductive toxicity. Still other tests evaluate specific damage to nerves, the immune system, and the hormonal system.

These exhaustive series of toxicological evaluations (Table 12.1) cost a fortune. The public should take considerable comfort in the fact that a repellent product passed these tests. National governmental registration authorities generally use teams of toxicologists to examine the data provided by manufacturers and testing laboratories. As the data come in, these experts compare probable exposure patterns to toxicological tests, also considering the results from similar molecules. If a result is acceptable for public safety, then the chemical may remain in consideration. Plenty don't pan out—a painful result for those who may have spent millions of dollars finding out that the chemical was not useful because of its toxicological properties.

GOVERNMENTAL REGULATION

In the opinion of the authors of this book, repellents can be divided into two categories: those that are government regulated and those that should not be used. Although you certainly have the right to doubt the safety of any product, using one that has never been authoritatively tested is a foolish risk. Animal testing is highly objectionable to many people, but it is difficult to see how the safety of a product could be assured without the use of laboratory animals. Those who perform this testing under modern conditions make every effort to treat animals in a humane manner, and they conform to rigid standards monitored by external boards known as institutional animal care and use committees.

Every country does not have to repeat all of the tests and evaluations of repellent products. The system of sharing data gets complicated, though, because the tests are expensive and companies want to maintain their exclusive rights to the information about their specific formulations. Most countries require some sort of approval before a particular product can be sold, but they might simply base their evaluation on another national government's approval. In fact, each of the states of the United States can accept, reject, or modify the approval by the EPA.

The major players in repellent registration are the World Health Organization, the United States, Canada, the European Union, Japan, and Australia. Each governmental authority does not necessarily agree on what is safe. For example, the United States rejects the use of pyrethrins and permethrin in topical repellents, but Australia accepts them. This is not to say that one country is right and the other is wrong. In this case, Australia has a different perspective from the United States on the threat from mosquitoes and the advantages of using topical pyrethroids. Until West Nile virus was introduced into the United States in 1999, mosquito bites were rarely infectious, which created a persistent attitude that a bite is more of a nuisance than a threat.[2] In contrast, Australia experiences thousands of cases of serious disease caused by two mosquito-borne viruses, Ross River virus and Barmah Forest virus. Northern Australia also continues to see people afflicted with the classical duo of Southeast Asian tropical, arthropod-borne diseases: scrub typhus (chigger-borne) and dengue. On the other side of the equation, the safety record of skin-applied pyrethroids is actually quite good, as observed through the use of louse and scabies mite treatments. American toxicologists worry that the long-term use of pyrethroids could result in skin sensitization or even poisoning, especially if an individual had simultaneous exposure to certain insecticides or drugs. Therefore, Australia has decided that the risk of toxicity is outweighed by the threat of mosquitoes, whereas the opposite judgment has been made in the United States.

2. However, during World War II, the United States suffered about a million cases of dengue, and prior to that, yellow fever and malaria had been periodic scourges of the country.

Table 12.2 Countries That Participate in the Organisation for Economic Co-Operation and Development's Pesticide Working Group

Australia	Hungary	Norway
Austria	Iceland	Poland
Belgium	Ireland	Portugal
Canada	Italy	Slovak Republic
Czech Republic	Japan	Spain
Denmark	Korea	Sweden
Finland	Luxembourg	Switzerland
France	Mexico	Turkey
Germany	Netherlands	United Kingdom
Greece	New Zealand	United States

The Organisation for Economic Co-operation and Development (OECD) is a nongovernmental organization that provides a medium for the harmonization of pesticide testing, including repellents. Although far from a stamp of approval, you can at least be sure that the safety of products labeled in one of the participating countries (Table 12.2) has been considered and probably thoroughly reviewed. Another valuable international resource is the World Health Organization Pesticide Evaluation Scheme. This organization systematically reviews the effectiveness and safety of public health pesticides, including repellents. Its aim is to provide an objective evaluation of potentially useful products for countries that might not otherwise have the resources to determine which chemicals are appropriate for their particular problems.

The regulation of the safety of products can impose many requirements on a manufacturer that hopes to sell a new product. If the active ingredients are well known, the company might only have to refer to those safety data, prove that its own product would not increase exposure over the levels in previous products, and go through a paper exercise to get an approved label. At the opposite extreme, a new active ingredient requires years of toxicological testing of the pure chemical and of formulations. The exact tests performed depend on what can be guessed from the chemical structure, especially compared to similar compounds. As data get submitted to the regulatory agency, there can be feedback on what kinds of further tests are required. As we have discussed, the toxicological assays themselves can be demanding, but in addition they must be performed in such a way that each procedural step and all materials are certified as accurate and genuine. The monetary investment builds up as each test is staffed, performed, and documented during a period that commonly stretches on for years. At any point, a result could show that the new active ingredient is too toxic for use as a repellent, making the investment to that point essentially worthless. This kind of product development is not for the fainthearted.

The final step in the regulation process is to write a label that states how the product should be used. In the United States, the pests are listed, but the specific requirements to prove effectiveness are still under development. The toxicological tests often result in safety considerations that are translated into specific restrictions on use. When a label says not to apply a product to children under a certain age, to keep the product out of the eyes, or to limit the number of applications per day, pay attention! Those instructions are based on millions of dollars' worth of tests, the considerations of responsible industrial and governmental specialists, and a desire to communicate with the person who eventually uses the product.

The determination of safety continues in a less organized way after the product is on the shelves. University or government scientists might perform epidemiological studies that evaluate the safety history of the product. Since repellents are intended to prevent bites, studies have also examined the medical benefits of their use. Another source of information is reporting from poison control centers or other medical reporting systems. Part of our confidence in the safety of DEET is based on the extreme rarity of negative effects from its application. For example, during 1993–1997 in the United States, 20,764 DEET applications that resulted in symptoms and intentional or accidental overdoses resulted in only 26 major poisonings and 2 deaths—a very low rate considering the use of DEET products by some 200 million people over those 5 years. In its much shorter history of use, picaridin has resulted in only one skin reaction. Most impressive, IR3535 has never caused any kind of poisoning during 25 years of product history. The U.S. Environmental Protection Agency uses these sorts of data and the accumulation of toxicological tests to re-review registrations. DEET's registration was renewed in 1998, as documented in a detailed publication available to the public.

So how do you know it is safe? The short answer is to use a product registered in a country that applies modern toxicological standards—*and* to follow the directions on the label. The big five active ingredients seem to be as safe as we can know. DEET and IR3535 have a long, good history. Picaridin has only been on the market for a few years, but it is already used widely with virtually no problems. What is more, picaridin is the first active ingredient developed under modern toxicological standards. PMD is considered safe and is widely used, but as a plant-derived compound, it has escaped some of the thorough testing of the synthetics. India has performed the full spectrum of safety tests on DEPA. Its military uses DEPA as a standard repellent, suggesting that its use history is satisfactory.

The picture gets murky for many other active ingredients, but that does not mean they are not safe. Some of the less common active ingredients you will see in a few registered products are pyrethrins, piperonyl butoxide, permethrin, the MGK synergists (MGK 264 and 326), and dimethyl phthalate (DMP or DIMP). The range of active ingredients will continue to change, but one thing will remain constant: it's cheaper to discover the chemical than to prove it is safe.

WILL IT HURT?

Toxic Stings and Bites

A person who gets bitten by a mosquito or stung by a bee puts the experience into a mental folder labeled something like "ick" or "painful" or "unfair—I was minding my own business." From the bug's point of view, there is a sharp dividing line between biting to suck blood and biting or stinging to defend itself. The blood suckers may have painful bites, but the point of the exercise is to get nutrition out of your body, not to inflict pain. It's never pleasant for us to make such a donation, but the pain is rarely sufficient to cause us to drop a hammer on a toe or fall off a ladder. On the other hand, bites from centipedes and stings from scorpions, wasps, ants, or bees are usually inflicted on humans in perceived defense—either of the individual bug or of a colony. The venoms injected by these arthropods often originated as part of the feeding or prey-handling process, but have evolved to a defense that is sometimes chemically modified to inflict lots and lots of pain. In some cases, the venom does its job of defense, followed by secondary damage associated with what would normally be the feeding process. Spiders are a little different in that they clearly bite humans in perceived defense, but the bites of all but the largest species are only mildly painful. The venoms of spiders vary a great deal in their toxicity for humans, as discussed below.

As pointed out in the second chapter, many people want to know why there are mosquitoes. We jocularly stated that our favorite answer is, "because there can be." When people ask this question, they actually expect to get an explanation of how mosquitoes are some sort of essential link in the world ecosystem. The fact is that mosquitoes and other blood suckers can be eliminated from entire areas with no obvious ecological effect other than an improvement in human health. On the other hand, the venomous groups of arthropods have a tremendous influence on ecosystems wherever they occur—and they occur everywhere except Antarctica. Their main influences are as predators (centipedes, scorpions, spiders, ants, and wasps) and as pollinators (bees, wasps, and the odd ant). Ants are so numerous that they also have major effects on soil structure and nutrient recycling. Simply viewed from the human perspective, these arthropods are not only beneficial, they are essential to human life as we know it. Without them, other insects would overrun the planet, many important flowering plants would cease to yield fruit and seed, and natural detritus would tie up nutrients. When we attempt to control blood-sucking bugs, we take care to protect the environment while coming close to eliminating the bugs. Controlling the venomous arthropods is more of a problem because the bugs themselves are important to maintaining a healthy environment for humans. Wholesale destruction is seldom justified, and the emphasis should be on avoiding the bites and stings. Insecticides should be used sparingly and only in situations where toxic bites and stings are a consistent problem.

Chapter 1 gives you all the tools you need to determine the general category of a bug that penetrated your skin to take blood or inject venom. This chapter will review each group of venomous arthropods, describing the dangers from their bites and stings, their distribution, how to keep them out of your house, and how to avoid them. Compared to the blood suckers, venomous bugs are a minor health problem—unless, of course, you are the one getting stung.

SCORPIONS

Almost everybody knows a scorpion when they see one. The name and the image pepper various cultures on everything from artwork to energy drinks, usually as a symbol of the insidious, strong, or dangerous. Physically, we know they have a crab-like appearance terminating in a flexible, stinging tail. There is nothing ambiguous about the impression they give.

Their alert stance on eight legs with upraised pincers in front and a springy tail poised to deliver a venomous blow make a clear visual statement of the ultimate, focused predator. No wonder they are the stuff of nightmares and taboos. Scorpions are also interesting creatures that live in a wide range of habitats in temperate and tropical parts of the entire planet. Among the fascinating scorpion facts are that they glow brightly under ultraviolet light (even some fossils retain this quality), and they give birth to live young that spend their first stage on the mother's back.

The scorpion biological lineage is ancient. They are possibly the oldest animal inhabitant of the land. By the Devonian era (350 million years ago), creatures that resembled modern scorpions crawled out of the primordial seas long before insects even existed. The largest fossil scorpion found so far is *Brontoscorpio*, which measures over three feet (0.98 m) long.

Spiders, ticks, and mites are related to scorpions and form the group of eight-legged arthropods known as arachnids. The greatest resemblance among these creatures is in the mouthparts. Both spiders and scorpions do a lot of digesting before the prey goes past the mouth. The sting venom and the saliva can start the digestion process of scorpions. The other similarity between scorpions and spiders is that they are all, without exception, predators. Scorpions eat whatever they can kill, tear apart, predigest, and stuff in their mouths. The menu often includes insects, but might also have mice, birds, spiders, or other scorpions. The predatory habits of scorpions make them beneficial to humans and the environment. In some habitats, they are probably significant natural regulators of their prey populations.

The damage from scorpions comes from the venom injected by their stings. Even the smallest scorpions (1/2″ or 1 cm, including the tail) have a stinger that is larger than the elegant hypodermic of the mosquito's drinking tube. Some of the larger scorpions—up to 7 1/4″ (20 cm)—can really pack a wallop, though the stings of the largest species are not medically dangerous. The scorpion drives that stinger home with the leverage of almost half its body length, using the muscles within the abdominal segments that have been modified into the iconic tail. In at least some species, the rapid and hard-driving sting is more typical of defensive than predatory behavior. The bottom line is that few scorpion stings are going to go unnoticed and that's exactly the scorpion's objective. Few who have experienced the searing pain and disturbing sense of danger caused by an unexpected zap from a scorpion will forget that they need to give these creatures a wide berth.

The initial pain, reddening, and swelling that get our immediate attention may last only a few minutes or as long as several hours. The majority of the more than 1,500 species worldwide cause only this kind of injury. Admittedly, almost any scorpion sting is going to be considered a significant inconvenience, but in reality, the effects will be no worse than a mosquito bite and without the danger of associated infection. Localized effects from the venom can be more disturbing, though still not dangerous. The swelling can persist several days, blood-filled blisters may form where cells have been damaged, local nerves may cause spasmodic twitching near the site of the sting, and some skin may slough off.

The venoms of about 50 species cause much more serious health problems, and half of those species have been responsible for deaths. Given the wrong circumstances, the number of deaths from scorpion stings is significant. For example, Mexico recorded almost 400 deaths and some 20,000 stings during the 1940s. Some current estimates go as high as 200,000 stings per year and 800 deaths in Mexico. The effects from these more dangerous venoms are called *systemic* because their damage extends beyond the site of the sting to other parts of the body. Scorpion venoms are complex mixtures of small proteins that, like snake venoms, have specific effects on tissues. The exact nature of the venom mixture varies among scorpions and sometimes between scorpions of the same species in different regions. Table 13.1 gives you some idea of the potency of the worst scorpion venoms.

Systemic scorpion venoms usually affect the nervous system, though the Middle Eastern thin-tailed scorpion (*Hemiscorpius lepturus*) is an exception. The sting from this scorpion causes massive tissue destruction similar to a rattlesnake bite, sometimes resulting in death by renal failure or other complications. The effects of deadly scorpions on the nervous system

Table 13.1 Toxicity of the Venoms of the Most Toxic Scorpions

From Mullen and Durden 2002		
Common Name	**Scientific Name**	**Amount of Injected Venom to Kill an Adult***
deathstalker scorpion	*Leiurus quinquestriatus*	0.0006 oz. (17 mg)
Moroccan fat-tailed scorpion	*Androctonus mauretanicus*	0.0008 oz. (21 mg)
yellow fat-tailed scorpion	*Androctonus australis*	0.0008 oz. (22 mg)
Arabian fat-tailed scorpion	*Androctonus crassicauda*	0.0010 oz. (27 mg)
Brazilian yellow scorpion	*Tityus serrulatus*	0.0010 oz. (29 mg)
alacrán de Morelos	*Centruroides limpidus*	0.0017 oz. (47 mg)
Egyptian yellow scorpion	*Androctonus amoreuxi*	0.0018 oz. (51 mg)
common European scorpion	*Buthus occitanus*	0.0022 oz. (61 mg)
Arizona bark scorpion	*Centruroides exilicauda*	0.0022 oz. (76 mg)
South African fat-tailed scorpion	*Parabuthus transvaalicus*	0.0103 oz. (289 mg)

*As discussed in chapter 12, toxicologists commonly gauge toxicity as the amount of poison necessary to kill half of a population of model animals, in this case injected into mice. The figures in the table were based on extrapolating the amount of injected toxin necessary to kill a mouse to the amount of toxin necessary to kill a 150-pound (68 kg) human.

cause a number of signs and symptoms. Initial pain at the site of the sting can change to numbness or be an ache surrounded by numbness. The lack of sensation can spread far from the sting site in conjunction with other signs of nerve involvement like slurred speech, difficulty breathing, irregular heartbeat, and nausea. Powerful toxins affect nerve transmission directly. The damage causes the release of chemicals that create a cascade of effects in the body. Profuse sweating, restlessness, priapis, salivation, and other signs are reminiscent of some kinds of insecticide poisoning. Even severe poisonings seldom end in death, though fatalities are much more common in children. Usually, a patient will survive if he is still alive after 12 hours, though deaths have been recorded anytime from 1.5 to 30 hours after the sting. The most common causes of death are cardiac or respiratory failure, sometimes associated with diffuse coagulation and leaky capillaries (that is, your blood stops flowing properly).

Unfortunately, it is not possible to give much guidance on how to tell a dangerous scorpion from a harmless one. All but one of the medically important scorpions belong to a single family called Buthidae, but this family also includes many innocuous species. All of the buthids have slender pincers, though this is hardly a conspicuous warning sign. The one exception to the slender-pincers rule is the dangerous Middle Eastern thin-tailed scorpion. It has rather large pincers and a thin tail. The color patterns and sizes of dangerous species are highly variable. They have longitudinal stripes or not, transverse stripes or not, light-colored bodies or dark-colored bodies (often in the same species), and range in size from less than 2″ (5 cm) to 6″ (15 cm) in length. The best advice is to adjust your level of concern with the region in which you travel or live, because large parts of the world do not have any really dangerous scorpions.

Table 13.2 lists species of scorpions generally acknowledged as medically important. People have had a serious reaction to species not on this list, and the list includes some species that have been associated with serious envenomization only a few times. Knowing the scientific names can help you to find more information, especially from local sources that may have detailed insights on the seriousness of the problem and what to do about it. Common names of scorpions have not been developed well, which will cause a certain amount of confusion between species.

Among these species are some infamously troublesome pests. Let's start with the Arizona bark scorpion, *Centruroides exilicauda* (sometimes called *Centruroides sculpturatus*). This little scorpion lives in the southwestern United States (Nevada, Utah, southeastern California, Arizona, and New Mexico) and northwestern Mexico. Mature specimens are only 2.5″ (7 cm) long and have the yellowish-brown color of dark straw. They may or may not have two darker stripes running along the length of the top side of their bodies. Those markings cause the Arizona bark scorpion to resemble the harmless and widespread common striped scorpion (*Centruroides vittatus*). You can tell the difference by looking for a dark triangular patch just behind the head, which is only present on the common striped scorpion. Unfortunately, the Arizona bark scorpion likes to live in houses, as well as in firewood, construction debris, and other items commonly handled by people. This species is reported to sting readily when disturbed, but only about 100 stings per year are reported in Arizona. There is less than one death per year caused by this species.

Mexico is home to a series of dangerous, related species of bark scorpions. They range in size from the alacrán de Nayarit

Table 13.2 Some of the World's Dangerous Scorpions

Common Name	Scientific Name	Distribution	Comments
The Americas			
Arizona bark scorpion	*Centruroides exilicauda*	Utah, Nevada, SE California, Arizona, New Mexico, NE Mexico	deaths in children; stings without warning; crevices and in houses
alacrán de la costa de Jalisco	*Centruroides elegans*	Mexico	dangerous venom
alacrán de Michoacán	*Centruroides infamatus*	Mexico	dangerous venom
alacrán de Morelos, de Colima, or de Iguala	*Centruroides limpidus*	Mexico	large venom yield; crevices; crumbling walls; mass stingings
alacrán de Nayarit	*Centruroides noxius*	Mexico	dangerous venom; less than 2 inches (5 cm) long
alacrán de Durango	*Centruroides suffusus*	Mexico	dangerous venom
Trinidad scorpion	*Tityus trinitatis*	Trinidad, Venezuela	dangerous venom; causes pancreatitis; coconut groves and cane fields
Amazonian scorpion	*Tityus cambridgei*	Amazon basin, northern South America to Panama	systemic toxic effects may last days; inhabits forests
escorpião do nordeste	*Tityus stigmurus*	northeastern Brazil	dangerous venom; hot dry places
Brazilian yellow scorpion	*Tityus serrulatus*	southern Brazil	most dangerous Brazilian scorpion; causes pancreatitis; infests houses, hot dry areas; usually parthenogenic
brown scorpion	*Tityus bahiensis*	southern Brazil, northern Argentina	dangerous venom; infests houses, wet areas
Argentinian scorpion	*Tityus trivittatus*	Argentina, Paraguay, Uruguay, Brazil	systemic effects but temporary; infests houses, wet areas
Mediterranean, North Africa, Middle East, Asia			
Egyptian yellow scorpion	*Androctonus amoreuxi*	North Africa, Middle East	one of the most dangerous scorpions
yellow fat-tailed scorpion, Omdurman	*Androctonus australis*	North Africa, Asia	dangerous venom; frequent association with people; most dangerous scorpion; under rocks, crevices, in buildings
black fat-tailed scorpion	*Androctonus bicolor*	Egypt, Israel, Jordan	medically important; dry habitats; under rocks
Arabian fat-tailed scorpion	*Androctonus crassicauda*	Turkey, Middle East, North Africa	one of the most dangerous scorpions
Moroccan scorpion	*Androctonus mauretanicus*	Morocco	one of the most toxic scorpions
common European scorpion	*Buthus occitanus*	North Africa, Middle East, southern Europe	toxicity varies through range

Middle Eastern thin-tailed scorpion	*Hemiscorpius lepturus*	Iran, Iraq, Pakistan, Yemen	unusual cytotoxic venom; only non-buthid dangerous scorpion; dry habitats; in damaged homes
threatening scorpion	*Hottentotta minax*	north central and eastern Africa	venom little studied but some deaths; dry places; in houses
deathstalker scorpion	*Leiurus quinquestriatus*	Turkey, Middle East, North Africa	very toxic venom; one of the most dangerous scorpions; under rocks or in burrows; dry habitats; not usually in houses
lesser Asian scorpion	*Mesobuthus eupeus*	southwestern and central Asia	serious effects rare; neuro- and cytotoxic venom; climbs well; dry habitats; not usually in houses
Mediterranean checkered scorpion	*Mesobuthus gibbosus*	Turkey	medically important
Iranian yellow scorpion	*Odontobuthus doriae*	Turkey, Middle East, Afghanistan, Pakistan	relatively mild venom; dry habitats; sometimes in houses
Blacktail thicktip scorpion	*Parabuthus leiosoma*	Egypt, eastern Africa, Arabian peninsula	sometimes systemic effects; arid and semi-arid habitats; under rocks, crevices
Southern Africa			
granulated fat-tailed scorpion	*Parabuthus granulatus*	southern Africa	most dangerous scorpion in southern Africa; arid habitats; crevices, in buildings
South Africa fat-tailed scorpion	*Parabuthus transvaalicus*	southern Africa	dangerous venom; can spray up to 1 m; sandy habitats
three-scraper scorpion	*Parabuthus triradulatus*	South Africa	medically important
fuzzy scorpion	*Parabuthus villosus*	South Africa	medically important
India			
Indian red scorpion	*Hottentotta tamulus*	India	most dangerous scorpion in the region; moist, cooler sites; climbs

(*Centruroides noxius*), which is less than 2″ (5 cm), to the alacrán de Durango (*Centruroides suffusus*), which grows to a length of almost 3.5″ (9 cm). Species can be dark or light colored and with or without up to four body-length stripes. They inhabit crevices, crumbling walls, and houses, placing them in close contact with the population. Tragically, up to three-quarters of the deaths from scorpion stings are of children. In the twenty-first century, the application of antivenins (preparations of animal antibodies injected to neutralize the venom) has been credited with sharp reductions in mortality. Scorpion stings are considered a significant public health problem in much of Mexico.

Rather unexpectedly, Central America experiences few problems with scorpions other than inconvenient stings that are more of a nuisance than a health threat. South America and the Caribbean islands are home to a group of about 100 scorpion species that are all considered dangerous to one extent or another. Chief among them is the Brazilian yellow scorpion (escorpião amarelo), *Tityus serrulatus*. As an example of its potential effect on public health, the state of Minas Gerais alone recorded 6,018 stings during 1987–1989 with 92 deaths. Its venom is unusual among all poisonous animals in producing inflammation of the pancreas, as well as the devastating nerve systemic effects typical of scorpion stings. The Brazilian yellow scorpion likes to live near people in houses, basements, and backyards. Populations of the pest are expanding in southern Brazil, especially in cities, where some communities consider the threat from this scorpion sufficient to organize neighborhood campaigns against it. The brown scorpion, *Tityus bahiensis*, is considered the second most troublesome scorpion in Brazil. Nearby Argentina has a scorpion problem that is less severe than in southern Brazil, but two species are recognized as medically important: the brown scorpion and another species called the Argentinian scorpion (*Tityus trivittatus*). The more tropical areas of South America are home to the Amazonian scorpion (*Tityus cambridgei*), which thrives in well-watered, forested areas. Human contact is not so frequent with this species, but its sting can cause systemic effects that last for days. Finally, *Tityus trinitatis* creates a problem in the cane fields and coconut groves of Trinidad and Venezuela. People working on those farms run a risk of stings that cause serious envenomization similar to that from the Brazilian yellow scorpion.

The common European scorpion, *Buthus occitanus*, occurs along the Mediterranean in southern Europe. This species also lives in North Africa and the Middle East, but it is the only dangerous species in Europe. The toxicity of the venom varies throughout its range, as does its coloration. It has particularly strong ridges on its tail and the typically narrow pincers of most dangerous scorpions.

Once you slip around the Mediterranean to the Middle East and North Africa, the variety of medically important species increases. Eight different groups of scorpions include dangerous species, creating problems in many different kinds of habitats. The appropriately named deathstalker scorpion (*Leiurus quinquestriatus*) is a very dangerous scorpion that tends to sting people in remote areas far from medical help. They vary in color from yellow to orange to dark red, and sometimes the segment of the tail just before the stinger is darker than the rest of the body. They are medium-sized scorpions, up to about 3.5″ (9 cm) long. Another species in the region is probably even more dangerous because it delivers relatively large quantities of venom. The yellow fat-tailed, or Omdurman, scorpion (*Androctonus australis*) varies in coloration but usually has a darker end to its tail. The ridges on the tail are particularly well developed. Stings from this species can be common in some communities. A related species in Morocco (*Androctonus mauretanicus*) completes the trio of the most dangerous scorpions of the North African and Middle Eastern regions.

The two other regions of the world with scorpion problems are India and southern Africa. The Indian red scorpion (*Hottentotta tamulus*) occurs in various color forms throughout the country and occasionally causes deaths. The two most dangerous species in southern Africa are the South African fat-tailed scorpion (*Parabuthus transvaalicus*) and the granulated fat-tailed scorpion (*Parabuthus granulatus*). The South African fat-tailed scorpion is considered less dangerous because it sticks to wild habitats. It has the interesting capacity to spray its venom up to one meter. The granulated fat-tailed scorpion is the biggest problem because it likes to live in houses.

Among the challenges of daily life, a scorpion sting is definitely one of the ones you'd rather prevent in the first place. As we discussed in chapter 2, one of the ways to prevent a bug problem is to place yourself where the pests aren't. Regionally, we have seen that the risk of a dangerous sting is much greater in some places than in others. Some of the worst areas are the southwestern United States, the drier parts of Mexico, Brazil, the northern and southern ends of Africa, the Middle East, and India. If you go to or live in one of these areas, pay attention to what local people say about the scorpion hazard and do some research about your particular location. Some places have many scorpions that live close to people. Although such situations are the exception rather than the rule, those who live there must make scorpion avoidance a part of their daily routine. If you have a choice, choose a town or village that doesn't have the problem.

When it comes to temporary visits to scorpion areas, there is traveling and then there is traveling. If you stay in the Hilton, see the museums, and generally stick to well-worn tourist paths, you are unlikely to even see a scorpion. On the other hand, if you go to fight in a war, perform field research, or indulge in ecotourism, then you must be a bit more careful. First of all, never sleep on the ground. Scorpions out on the hunt might find your sleeping bag or rumpled bedclothes an attractive, safe resting site. Even the short legs of a military cot are enough to discourage most of them from climbing up, though the protection is ruined if a blanket drapes from the edge or an object leans against the cot. Use a bed net for further protection, especially from scorpions that might drop onto your bed from above. When you get up at night, watch where you put your feet and other tender parts of your body; a flashlight is a good friend when scorpions are about. If scorpions are common in the area, it may be worth investing in an LED (light-emitting diode) ultraviolet flashlight, which will clearly reveal the fluorescence (usually greenish) of any scorpion. Most people keep boots, shoes, clothing, and towels outside of their cots at night. Be sure to shake out these items in the morning. During the day, it is a good idea to put all that stuff on your cot under the bed net. Be mindful of where you put your feet and hands day or night. For example, if the task is something like taking down an old adobe wall or loading firewood, do not handle pieces unless you can see what is there. Even better, wear stout leather gloves and boots during such work. Children are more likely to be stung, and the effects of the venom on them are often much worse. It might not be such a great idea to take a child on a recreational trip to a place with abundant dangerous scorpions. If you must, then make sure the child understands the hazard, watch where the child goes, and examine clothing and towels for the child. It's hard to be vigilant all the time. The senior author of this volume was stung while packing a duffle bag in the middle of the day, even though he was well aware of the abundant scorpions in the Mosquitia, a remote region of Honduras. As with any adventure travel, it's prudent to know exactly where there is medical help and how someone can get you to it.

It is easier to avoid scorpion stings when you are actually living in the area. For one thing, you learn where they are and how to avoid them. A householder can do a lot to reduce the number of scorpions on her property and by building correctly in the first place. Dark empty spaces below the house, such as basements and crawl spaces, are attractive to some scorpions. Floors at the two extremes of design are the best for making the house more habitable for humans than for scorpions. A brick, tile, or cement floor laid directly on an earthen pad eliminates potential hiding places under the floor. It is important, however, to elevate the house by at least one step to prevent scorpions from walking straight in. Glazed tiles embedded in a line on the vertical face of the step will discourage scorpions from attempting to go up. At the other extreme, scorpions have a hard time getting into an elevated home. The space under the house should be tall enough (usually at least 9 feet [3 m]) to allow regular, thorough cleaning and organization of the space, denying scorpions extra places to hide. The lower 3 feet of the sides of the house should be as smooth as possible. Planed cement is often recommended as a suitable coating. Keeping trees and shrubs trimmed away from the house is particularly important. Other recommendations for making a bug-proof house are in chapter 4, and those measures will also help to prevent the entry of scorpions. Outside the house, cleanliness is next to scorpionlessness. Scorpions love the structure of debris and trash for hiding, in part because these same places tend to have more insects and rodents for them to eat.

The application of pesticides to control scorpions is seldom necessary and often undesirable. In general, it takes a large volume of material to treat all the places where scorpions might occur. Such treatment is expensive and has an impact on the local environment. Also, remember that scorpions are definitely beneficial in that they eat quantities of insects that we consider to be pests. Scorpion pesticide treatment is probably justified under some circumstances, however. For example, some towns in Mexico and Brazil were so plagued with dangerous scorpions that they launched campaigns to control them using both pesticide treatment and housing modification. Some species of scorpions can establish a true infestation in a home, increasing to numbers that make stings inevitable. Some structural modification might be necessary to prevent reentry, but such a situation would demand prompt pesticide application, especially if children are in the house. Pesticide applications would also be a wise idea before establishing a large-scale military bivouac or civilian campsite in a location with abundant scorpions. Although such applications may be harsh environmentally, the prevention of dozens or even hundreds of stings justifies the use of insecticides. Most of the same active ingredients mentioned in chapter 5 are also effective against scorpions. It is easy to find labels that list scorpions, and only those products should be used for the purpose. An effective treatment is likely to require maximum rates and broad bands of application in order to be sure that these large arthropods absorb a lethal dose. Some sources claim that dusts, rather than sprays, are more effective because the dust accumulates on the scorpion as it walks along close to the ground. The scorpion also may ingest the dust during grooming.

If you are stung by a scorpion, your first impulse may be to exercise your vocabulary of expletives. That is perfectly understandable, because a sting happens suddenly and often hurts like crazy. After the first shock, remain calm and get someone to help you. In areas where dangerous scorpions do not occur, you may not need medical attention at all. On the other hand, if you are in one of the places where medically important scorpions are abundant, it would be wise to get medical assistance as soon as possible. Poison control centers can offer good advice. Any scorpion sting should be washed thoroughly, and you should get a tetanus shot if you haven't had one recently enough. In general, try not to move around too much after a sting, keep the affected limb elevated, and apply a cold compress. If the scorpion was killed, take it with you to the medical center so that the doctors have a chance of using the appropriate antivenin, but do not take the time (or the risk) to capture a live scorpion.

SPIDERS

Scorpions may be a common symbol in many cultures, but spiders are a part of the human soul. If you are alert, it is difficult to go even part of a day without seeing, hearing, or feeling something to do with spiders. Every reference to a web harks back to the silken snares built by many species of spiders. Children's songs and books often feature humanized spiders, something that seems natural to most of us. Household cleaning includes the removal of webs in the corners of rooms and sweeping up the cast skins of the insects killed and eaten by spiders. It would be a rare house that did not contain spiders hidden away behind furniture, under woodwork, or within the walls themselves. Outdoors, we can see spiders everywhere if we pay attention and feel their webs break as we walk down the street.

By some counts, there are about 40,000 described species of spiders in the world. Some species are widely distributed because we have carried them about in our possessions, even to Antarctica (though none have successfully become established there). Although impressive, the number of species probably underestimates the importance of spiders in the world's ecology. Spiders have many adaptations and some are very specialized, but as a group they are extremely adaptable predators. Physiologically, they seem to be able to thrive on a great variety of prey, and some of them can wait for up to two years between meals. Behaviorally, many species will alter their hunting strategy according to the abundance of particular sources of food. Just to give a household example, house spiders eat about one mosquito per day when the insects are available, a significant dent in the mosquito population when you figure that a tropical home can contain dozens of these spiders. In an extreme example in the Gulf of California, spiders fueled by detritus-breeding flies attain a density of 200 spiders per cubic meter. Just imagine what the world must look like to a small fly faced with such an array of hungry predators. It is probably not an overstatement to say that human life is dependent on the way spiders keep other arthropod populations in check.

Unfortunately for us, about 200 species of spiders are said to be capable of inflicting toxic bites on people. There are other spiders with highly toxic venoms, but unlike scorpions, the venom is delivered by the mouthparts (technically, fangs attached to the chelicerae), and those mouthparts are much more adapted to feeding on their prey than to delivering defensive bites to large mammals. As a result, the vast majority of spiders have fangs so small that they would have a hard time penetrating the outer layers of our skin.

Camel spiders (other common names are sun spiders and wind scorpions) are arachnids that look a lot like spiders, but they are actually in a different group. They are impressive-looking animals with huge jaws held prominently in front of their heads, and they are capable of running very fast. These amazing arthropods are common in dry, hot areas like the Middle East and the American Southwest, though they range as far north as the state of North Dakota in the United States. Camel spiders can inflict a real laceration, but they are not venomous. The bites occur rarely, for example, when someone is careless about picking up a camel spider or when one of them is trapped between a person and clothing or sleeping bag. The awesome appearance of camel spiders makes them the object of some pretty ridiculous myths, but they are basically harmless creatures.

Table 13.3 summarizes the kinds of medically important spiders, where they occur, and generally how their toxins affect people. As you can imagine, no one does experiments on people to determine how a spider bite affects them. The information is mostly based on anecdotal experience, but documented experience is rare because people rarely capture the spider that bit them. Experts have done their best to estimate and generalize, which accounts for the use of some broad categories and vague comments. Related species not mentioned in the table could have similar venoms, but those species may be less important because of low

Table 13.3 A Summary of Medically Important Spiders

Common Name	Family	Scientific Name	Distribution	Toxicity
Most Dangerous				
Sydney funnel-web spider	Hexathelidae	*Atrax robustus*	Sydney metropolitan area, southeastern Australian seaboard	funnel-web spiders: powerful neurotoxins that cause life-threatening shutdown within 15 minutes followed by more slowly expressed pulmonary and circulatory problems
North Coast or tree funnel-web spider	Hexathelidae	*Hadronyche formidabilis*	coastal southeastern Queensland and northern New South Wales, Australia	
southern tree funnel-web spider	Hexathelide	*Hadronyche cerberea*	coastal southern New South Wales, Australia	
banana spider, armed-banana spider	Ctenidae	*Phoneutria nigriventer*	Brazil, Uruguay, Argentina	severe neurotoxic effects from peptide and nonpeptide stimulants of nerve activity, severely painful, followed by systemic effects; seldom fatal
southern black widow spider	Theridiidae	*Latrodectus mactans*	North and South America, Hawaii, Saudi Arabia, India, Australasia, Pacific islands	widow spiders: 30 minutes to a few hours after moderately painful bite, severe nervous systemic effects caused by massive release of neurotransmitters, including generalized pain and muscle spasms, nervous grimace, abdominal rigidity; lasts 3–7 days; sometimes fatal
northern black widow spider	Theridiidae	*Latrodectus variolus*	North America	
western black widow spider	Theridiidae	*Latrodectus hesperus*	Western United States; introduced into Israel	
araña del trigo	Theridiidae	*Latrodectus curacaviensis*	South America	
European or Mediterranean black widow spider, la malmignatte, karakurt	Theridiidae	*Latrodectus tredecimguttatus*	Mediterranean region and southern Russia	
black button spider	Theridiidae	*Latrodectus indistinctus*	South Africa	
Australian red-back spider	Theridiidae	*Latrodectus hasselti*	Australia	
katipo spider, New Zealand red-back spider	Theridiidae	*Latrodectus katipo*	New Zealand, Caribbean	

(Continued)

 Table 13.3 *(Continued)*

Common Name	Family	Scientific Name	Distribution	Toxicity
brown recluse spider	Sicariidae	*Loxosceles reclusa*	southeastern and central United States	recluse spiders: genus has 100 species worldwide with potentially similar toxins; relatively painless bite followed by blistering, reddening, and (rarely) bull's-eye pattern; black eschar forms in middle and expands with gravity, creating wound that takes months to heal and may reach to muscle layer; occasionally systemic effects 2–7 days after bite with hematological, respiratory, and circulatory problems
South American violin spider	Sicariidae	*Loxosceles laeta* (also *gaucho*, *intermedia*, and others)	South America	
Mediterranean recluse spider	Sicariidae	*Loxosceles rufescens*	Mediterranean region, eastern United States	
six-eyed sand spider	Sicariidae	*Sicarius hahni*	South Africa (other species throughout Southern Hemisphere)	
Medically Important				
hobo spider	Agelenidae	*Tegenaria agrestis*	Europe, introduced into northwestern North America (Utah to Alaska)	bites have not been proven to be dangerous; probable that perception of medical importance is false
agrarian sac spider, yellow sac spider	Miturgidae	*Cheiracanthium inclusum*	Western Hemisphere	sac spiders: most common of spider bites; immediate local pain followed by wheal similar to wasp sting, numbness, small skin lesion, sometimes mild systemic effects; probably never dangerous
long-legged sac spider	Miturgidae	*Cheiracanthium mildei*	Europe, introduced into Western Hemisphere	
Dornfinger (thorn finger) spider	Miturgidae	*Cheiracanthium punctorium*	Europe, Central Asia	
yellow night-stalking sac spider, pale leaf spider	Miturgidae	*Cheiracanthium mordax*	Australia, introduced into Hawaii and Fiji	
Japanese sac spider	Miturgidae	*Cheiracanthium japonicum*	Japan	
South African sac spider	Miturgidae	*Cheiracanthium lawrencei*	South Africa	

false black widow spider, cupboard spider	Theridiidae	*Steatoda*	worldwide	toxin similar to that of widow spider, but bites much less dangerous
brown widow spider	Theridiidae	*Latrodectus geometricus*	Brazil, introduced worldwide in tropics	considered least dangerous of widow spiders; rarely bites; common in houses and yards
tarantulas (various colorful names from the pet trade)	Theraphosidae	*Acanthoscurria, Pamphobeteus, Phormictopus, Sercopelma*	Latin America	various toxic effects with moderately severe systemic nervous effects and tissue destruction; some tarantulas in other genera are harmless to humans
Chilean mouse spider	Actinopidae	*Missulena tussulena*	Chile	mouse spiders: painful bite; record of one serious bite from *M. bradleyi*; neurotoxic effects
eastern mouse spider	Actinopidae	*Missulena bradleyi*	eastern Australia	
red-headed mouse spider	Actinopidae	*Missulena occatoria*	Australia	
baboon spider	Theraphosidae	*Harpactirella lightfooti*	South Africa	bite causes intense pain followed by systemic effects ending in shock and collapse; recovery in 24 hours
tree spider	Theraphosidae	*Poecilotheria*	India, Sri Lanka	toxic venom similar to black widow spider, though no fatalities recorded
sheet-web or funnel-web tarantula	Dipluridae	*Trechona venosa, T. zebra*	South America	sometimes fatal
Nuisance Spiders				
labyrinth spider	Agelenidae	*Agelena labyrinthica*	Europe	nuisance spiders: typical signs are pain, swelling, and reddening; some individuals may have a worse reaction or allergic response
tangled nest spider	Amaurobiidae	*Coelotes obesus*	Europe	
Corinnid sac spider	Corinnidae	*Trachelas*	United States	
Florida false wolf spider	Ctenidae	*Ctenus captiosus*	Florida	
ground spider	Gnaphosidae	*Herpyllus blackwalli, H. ecclesiasticus*	United States	
white-tailed spider	Lamponidae	*Lampona cylindrata*	southeastern mainland Australia, Tasmania	

(Continued)

Table 13.3 *(Continued)*

Common Name	Family	Scientific Name	Distribution	Toxicity
wolf spider	Lycosidae	*Lycosa*	worldwide	
green lynx spider	Oxyopidae	*Peucetia viridans*	southern United States to Venezuela	
raft spider, fishing spider	Pisauridae	*Dolomedes fimbriatus*	Europe, northern Asia	
jumping spider	Salticidae	*Phidippus johnsoni* and a few other species	worldwide	
tube web spider, cellar spider	Segestriidae	*Segestria florentina*	Mediterranean region, introduced widely in Europe, Argentina, Australia	
crab spider	Thomisidae	*Misumenoides*	Western Hemisphere	
tarantula	Actinopodidae	*Actinopus*	South America	
brush-footed trapdoor spider	Barychelidae	*Idiommata blackwalli*	eastern and northern Australia	

abundance, distribution far from people, or less aggressive behavior.

DANGEROUS SPIDERS

Australian funnel-web spiders draw a lot of attention from people who live near them. They are in the same group of spiders as tarantulas, sharing a somewhat hairy appearance, strongly constructed bodies, and large size (up to 2.5″ [6 cm], including legs). The part of the body with the legs is shiny black and the abdomen is velvety, ranging in color from dark reddish to black. Three particular characteristics of these spiders are their large forward-pointing fangs, the flat top with a dimple in the middle as seen from above between the legs, and long spinnerets (where the silk comes from) at the end of the abdomen. The Sydney funnel-web spider is the most dangerous of the eight species known to bite people. It lives in silken tubes built under rocks and debris and is not shy about living in backyards or even in houses. The males are the biggest problem because they tend to wander about at night looking for food and mates. Combining an incredibly powerful set of nerve toxins, a readiness to bite defensively, and a tendency to live near people, these spiders are a true public health problem in the Sydney area and in an expanding part of Victoria and New South Wales. There have been 13 documented fatalities from these spiders' bites since 1926. The relatively low number is probably due in part to the availability since 1981 of effective antivenin, developed by Dr. Struan Sutherland of Commonwealth Serum Laboratories, Victoria, Australia. Not every bite results in severe poisoning, even without treatment.

The armed spiders of southeastern South America might be mistaken for tarantulas, but they are actually related to the larger group of spiders whose fangs work side-to-side rather than front-to-back. A number of species in this group bite people. The worst species is the armed-banana spider. It is large (up to 4″ [10 cm], including legs) with dull gray background coloration, red hairs around the fangs, and a light-colored stripe along the length of the top side of the abdomen. Although this description sounds distinctive, there is a harmless spider that lives in the same area that is very similar in appearance. When threatened, the armed-banana spiders raise their front two pairs of legs and face the source of their concern with fangs extended and leg hairs bristling. Armed-banana spiders are wandering nocturnal hunters that do not build webs to catch their prey. Unfortunately, they often in the early morning hours wander into homes, where sleepy inhabitants get bitten as they dress, roll over in bed, or dry off with a towel. This wandering habit makes bites alarmingly common—800 per year in the São Paulo, Brazil, area alone. The bite is extremely painful and is often followed by disturbing general signs of nervous stimulation. Fewer than 1 in 100 people die as a result of a bite, and most people fully recover after 48 hours. The toxin is reported to be complex and includes nonprotein components. Unlike other spider bites, first aid recommendations for armed spiders include the application of heat rather than cold to the bite site.

Widow, or shoe-button, spiders catch their prey in messy tangled webs, and for some reason they have developed one of the most powerful toxins known. Males are much smaller than females (1¾″ [4 cm] with legs) and are considered to be far less dangerous. Widow spiders usually live outdoors, coming inside sheltered areas like basements, garages, and even neglected areas of the home itself. Some locations, like southern California, have more than the usual density of widow spiders, though one species or another occurs almost everywhere in tropical and temperate climates. Female widow spiders of all types usually have large, globe-shaped abdomens, dark coloration, and some sort of red or orange markings on the bottom side of the abdomen (typically, hourglass-shaped in the southern black widow spider). Two exceptional color patterns occur on common widow spiders. The Australian redback spider has markings on the top of the abdomen (which is sometimes light-colored), and the brown widow spider of South Africa is often light-colored with a highly patterned abdomen. The sharply bent legs are slender and the part of the body holding the legs and head appears to be very small compared to the abdomen. Females usually hang upside down in their webs in typical spider fashion, often with an egg sac suspended somewhere nearby. It's an unlucky person who gets bitten by one of these shy spiders. The victim is in for as much as a week of painful, disturbing sensations as the toxin does its work on the nervous system. Good antivenin treatment can reduce this period to as little as an hour. The antivenin from one species often serves as an antidote to the bites of other widow spiders and even false widow spiders (genus *Steatoda*). Bites usually occur when people inadvertently press against the spiders while putting on clothing, sitting in an outdoor latrine, or working with plants.

Gory. Horrible. Unimaginable. The damage caused by recluse spider bites can dissolve a large part of the flesh and fat of a limb, exposing the muscle beneath. Photographs of victims (blessedly not reproduced here) not only turn your stomach, they also make you feel like you should kill every spider that ever existed. This would obviously be an overreaction to a group of spiders that actually bites infrequently and, for the most part, lives as a good neighbor with people. Most bites occur first thing in the morning, when people trap a spider between clothing, bedding, or towels and their bodies. The dangerous female recluse spider is usually dull yellowish to dark brown and about 1″ (3 cm) long, including its rather long legs. The only distinguishing marks occur on the front of their bodies where the legs come out. In some species, this marking resembles a violin with the neck facing the back of the spider. The arrangement of the eyes is also distinctive. Examination of a dead recluse spider with a hand lens will show that it has six eyes arranged in three pairs. They live throughout the tropics and temperate regions, bedding down in messy webs built among debris and other objects, both indoors and outdoors. An expanding dark sore surrounded by a reddened area, a pale ring, and a bluish ring (known as the "red, white, and blue target sign") often follows a bite. The South American species kill about 1.5% of people who are bitten, with serious systemic poisoning in 13% of cases. Antivenins are not necessarily helpful, cutting out the bite site does not stop the process, and drugs have achieved only mixed results. In North America, "only" about 3% of suspected bites require skin grafts. If you have symptoms that suggest you have been bitten by a recluse spider, be sure that your physician considers other, more likely causes, including bacterial or fungal infections, rickettsial disease, vascular disease, and exposure to irritant chemicals or poison ivy.

A word or two about tarantulas is important because these animals are commonly traded as pets. You will notice that the word *tarantula* is used for a number of spiders in Table 13.3. Technically, all spiders that move their fangs front-to-back (suborder Mygalomorphae), rather than side-to-side (suborder Araneomorphae), are in the tarantula group.[1] The southern United States is home to some very gentle tarantula species that are large and interesting creatures, the sight of which is usually an occasion for interest. Because the bite from these American tarantulas is mild—and they seldom bite anyway—there is

1. Sometimes, the word *tarantula* is applied to a large wolf spider in southern Europe that was the purported cause of *tarantism*, an ailment supposedly cured by incessant dancing. In fact, the illness was probably caused by the Mediterranean widow spider, sometimes called the *tarantola*. The phenomenon of tarantism that started in the fourteenth century produced a folk song and dance called the *tarantella*.

sometimes the impression that all tarantulas are harmless. In fact, when you move into Latin America, Africa, and Asia, some of these large, beautiful spiders have powerful venoms that affect the nervous system and that have tissue-destroying effects. Those who keep these creatures as pets or who encounter them in their native habitats should be careful to avoid bites. In addition, many of these spiders defend themselves by literally throwing tiny, barbed hairs. In the United States, the result can be irritation to the eyes, respiratory tract, and skin. The tarantulas in Latin America have more penetrating hairs that can cause long-lasting damage, especially to the eyes. You are only likely to encounter this problem if you keep tarantulas as pets, so put goggles or glass between you and the object of your affection.

MANAGEMENT AND AVOIDANCE OF MEDICALLY IMPORTANT SPIDERS

The vast majority of people do not need to do anything to reduce spider populations in their surroundings. They will never be bitten by a spider nor suffer a serious medical problem from a bite. Letting spiders do their thing is probably the best advice most of the time. Sometimes, when resting sites (that is, untidy possessions and trash) and prey items (that is, bugs that should be discouraged by following the advice in chapter 4) are abundant, the number of spiders increases so much that bites are bound to occur. Spiders are amazingly adept at building up their populations where there is a lot of food; therefore, limiting insect populations around the house can go a long way toward discouraging huge numbers of spiders.

Of course, spiders that commonly bite are unwelcome neighbors. Using seals on doors, placing screens on windows, and blocking holes that lead into the house will help to keep spiders out. Shrubs and trees should be trimmed away from houses. A tidy house with good sanitation provides spiders and their prey fewer places where they feel secure. Regular use of a broom to remove webs tends to discourage spiders, almost as though they learn to avoid those areas. Outdoors, you can wash down webs with a hose. Spiders are often the target of insecticidal treatment. Though harsh, insecticides can do a good job of clearing spiders out of an area for a surprisingly long time. As always, be sure that the label mentions use against spiders and that you observe all directions and precautions. There are also special spider-attracting sticky traps on the market, though they have never received thorough, independent testing for effectiveness.

Personal protection against spider bites involves many of the same steps described for scorpions. If you work in the garden, stack wood, or clean things up outdoors in an area with dangerous species, it may be worthwhile to wear gloves, boots, long sleeves, and long trousers. Brushing yourself off can prevent spiders from hitching a ride into your house. As you work, watch the materials you handle and the places you put your hands in order to avoid surprising a spider that would otherwise not care whether you are there or not. Night and early morning are the most likely times to be bitten because many spiders are nocturnal and because it gets difficult to spot them. Whether camping or in your own backyard, use a flashlight at night to see where you are going and what you are doing. Outdoor latrines are notorious as good locations for widow spiders, so inspect under the toilet seat before settling in. Inspect shoes, clothing, and towels before they come into contact with your body, especially in the morning. Children often suffer the most from spider bites; be sure they know how to recognize the dangerous species and check their surroundings carefully.

Don't panic if you are bitten by a spider. For the most part, the bite will only cause a little pain and perhaps some swelling and redness. If worse signs develop, seek medical attention right away, hopefully either with a description of the spider or possibly with the dead spider itself. You can get advice from a poison control center. Wash the bite site thoroughly to avoid secondary bacterial infection and get a tetanus shot if your previous vaccination has expired. Any systemic effects, like sweating, swelling of nearby lymph nodes, or other signs that the effects of the venom have gone beyond the bite site, should be cause to seek medical help. In those cases when the effects of the bite appear to be serious, elevate the bite site relative to your body, avoid activity, and apply cold compresses (warm for armed-banana spider bites) to the bite on your way to medical help. Funnel-web spider bites require unique, special treatment right away to avoid the possibility of rapid collapse. You should consult a medical professional about the procedure, but first aid for funnel-web spider bites involves placing a splint on the affected limb, wrapping it loosely in elastic or crepe bandages, keeping the victim immobile, and transporting the person to a center that has the antivenin.

CENTIPEDES

Centipedes have way too many legs. They have one pair of legs per segment, and they keep adding segments as they grow, producing up to an astounding 181 pairs of legs. The legs of the common house centipede are long, giving this frequent inhabitant of homes a distinctive appearance. Most centipedes have much shorter legs and a generally flexible, worm-like appearance (though unworm-like speed and agility). You see them whenever you dig around in the garden, turn things over in the garage, or move objects resting on damp soil. Most are predators that do their bit toward keeping the planet from being overrun with insects.

All centipedes have poison claws at the rear of the head. They use the claws to grasp their prey and inject a toxin. As with almost all venomous arthropods, the toxins immobilize and predigest the prey. Sometimes, centipedes bite people, usually when they are handled but occasionally when the centipede is simply crawling across someone who is sleeping. Some species are downright testy and bite quite readily. Bites are described as intensely painful, especially from the larger species. The toxin causes a burning sensation, itching, aching, and local swelling. Some bites may cause systemic effects, like headache, nausea, or vomiting, though they are rarely fatal. Signs and symptoms usually subside within 24 hours, and there is no specific treatment.

A wide variety of centipedes bites people, most often in warmer climates. The common eastern centipede (*Hemiscolopendra marginata*) is only about 2″ (5.7 cm) long. It lives in the eastern and southern United States. This species occurs in houses and supposedly bites without provocation. The red-headed, or giant desert, centipede (*Scolopendra heros*) occurs throughout the southern United States and Mexico. It commonly reaches a length of 6″ (15 cm) and delivers a painful bite.

Even its footsteps can leave tiny, toxin-laden punctures. A series of related species (subclass Scolopendromorpha) occurs worldwide and includes the largest centipedes. The Egyptian blue centipede (there are various other common names for *Scolopendra morsitans*) is about 5″ (13 cm) long. This species, which originated in Africa, now occurs in warm regions everywhere. It is a common cause of household bites in many parts of the world. Other notable species are the Australian giant centipede (*Ethmostigmus rubripes*), which attains a length of 10″ (25 cm), and the Amazonian giant centipede (*Scolopendra gigantea*), which is commonly 12″ (30 cm) long. These large centipedes are fearsome predators, leaping about to capture prey, including snakes and bats. Amazingly, the large centipedes will carry their paralyzed prey in their front legs, while walking on their hind legs.

Although centipedes occur almost everywhere and many species will inflict a mildly to wildly painful bite, they are seldom cited as a common household problem. Some countries, like Turkey, report many centipede bites, but the medical consequences are generally minor and the centipedes can be kept out of homes in most cases. It pays to check clothing, shoes, and towels in tropical areas and to be careful where you put your hands and bare feet. The precautions against scorpions and spiders should be helpful against centipedes, though the large species may climb onto cots and beds more readily. If centipedes are a common problem in your area, it might be wise to use a bed net just to prevent a painful wake-up call. The senior author remembers one soldier's description of a 10″ colorful centipede biting him on the chest in the morning; it's an image that certainly gets your attention!

WASPS

Wasps as a group have a tremendous impact on insect populations. Thousands of species of mostly tiny wasps specialize in parasitizing particular insects that we regard as pests. These wasps may not have any venom at all, using their stinger for its original purpose: to deposit eggs. Other wasps lay eggs through an opening at the base of their stingers, the stingers themselves being used primarily to inject venom either as a defense or as part of the prey-handling process. Males never sting because they do not have a modified egg-laying apparatus. Solitary wasps generally go about their business without stinging people, though some can give a painful sting when disturbed. The social wasps that live together in groups often mount organized defenses of their nests. Unfortunately, they can end up stinging people who are not even aware of the nest and have had no time to avoid the colony.

Stingers are not simple insect hypodermic needles. They consist of a sheath made up of several interlocking pieces. Inside the sheath are lancets that drill down into prey (if the wasp is procuring food) or tissue (if the wasp is defending itself or its nest). Venoms are primarily mixtures of proteins that paralyze, kill, digest, or specifically cause pain. The reaction is usually local, but some people develop an allergy to the stings. Allergies can vary from an uncomfortable amount of swelling to full-blown, fatal anaphylaxis. The amount of pain depends a lot on the species of wasp inflicting the sting.

The solitary wasps include three groups that commonly sting humans, though none are as troublesome as the social wasps. The velvet ants (family Mutillidae) have large, winged males and smaller (usually less than 0.5″ [1 cm] long) wingless females. These are beautifully colorful insects that sport flaming orange, ebony black, and metallic gold or silver coloration. Some people have dubbed the hairy California variety the "electric teddy bear." They have extremely painful stings—there's nothing cute about this teddy bear. Eight thousand species occur worldwide, especially in drier areas. The spider, or pepsis, wasps make a living by paralyzing spiders, which then receive an egg and serve as the nursery for the baby wasp. The famous tarantula hawk wasp of the western United States is huge, matching the size of its prey. It cuts a fine figure with its shiny blue-black body and bright orange wings. These wasps rarely sting, but when they do, the sting is considered to be one of the most painful. The mud daubers and cicada killers (family Sphecidae) form a diverse group of wasps that are not very inclined to sting. They often live in or around human dwellings in nests made out of mud. They provision the nests with caterpillars or other insect food items for their young. Making these nests involves a lot of trips back and forth.

Social wasps have colonies with one or more queens that do all the egg laying while the workers do the work. Part of that work is to defend the nest. It is made from a paper-like material produced by chewing up sources of cellulose, like wood and leaves. The nest can be located underground, in a wall, or hanging from a structure or tree. Perennial colonies of some species can get very large as generation after generation builds the structure to a size over 3 feet across (>1 m) and 6 feet (2 m) high. These kinds of wasps defend their nest vigorously, sometimes inflicting many stings on a person. They also sting when food supplies are scarce and the workers get aggressive about taking food scraps at picnics, around garbage cans, etc. Among the most common kinds of social wasps are the yellow jackets (genera *Dolichovespula* and *Vespula*), the hornets (*Vespa*), and the paper wasps (*Polistes* and *Mischocyttarus*).

It is possible to avoid the stings of velvet ants, spider wasps, mud daubers, and cicada killers simply by leaving them alone. It is a little harder with mud daubers because they can be numerous and often wander indoors, only to become stuck banging against the inside of a window. A person can get stung while trying to maneuver the wasp back outside. If you don't mind killing the wasp, a sharp direct hit from a fly swatter will crush these tough insects. Alternatively, most wasps are immobilized quickly by direct application of an indoor, aerosol insecticide.

The social wasps can be a real problem near a home. It is usually worthwhile to destroy the colony just to be able to go through your daily routine without stings. You can discourage new colonies with a broom or stick, but large ones may require insecticidal treatment. Commercial products that spray a large volume of insecticide solution for 10 or 20 feet (3–6 m) can be applied at night when most individuals of the colony are present and when they are less active. If things do not go well, feel free to retreat in an undignified manner. Large colonies, underground colonies, and, especially, colonies built inside homes really require professional pest control service. Commercial traps that you bait with either food or highly attractive chemicals can do a good job of cleaning out foraging yellow jackets that might sting during picnics or outdoor activities.

Popular lore suggests that bright clothing and floral-scented body care products encourage the approach of wasps, though there is only anecdotal evidence that this is the case. Some insect repellents discourage wasps, including IR3535.

ANTS

Ants built a large part of the earth on which we live, and if you weighed all the living things on our planet, as much as 15% of the weight would be ants. They are nearly everywhere except Antarctica, and they epitomize the industry of individuals working toward the accomplishments of the group. The humanized image of the self-sacrificing, hard-working ant toiling away without thanks for the good of the colony captures our imagination, even though it implies moral intentions that simply do not exist in insects. On the other hand, the ant colony as a whole is definitely the product of the coordinated efforts of its many members. Colonies range in size from a few workers to 300 million, and they all depend on communication and behavioral adaptations that result in survival of the group. A typical ant colony has biologically specialized individuals that may be physically adapted for defense, food gathering, care of offspring, food storage, or reproduction. Ants have an incredible variety of lifestyles among their 9,000 species. The typical pattern is a colony with thousands of individuals, including a long-lived, egg-laying queen (or several queens) fertilized by one or many males that have long since died. The queen does nothing but lay eggs, which are cared for by her nonfertile, female offspring. The offspring perform many tasks, feeding the queen and the developing larvae, bringing in food, and building the tunnels and chambers that make up the underground nest. Some species have specialized soldier ants with larger heads to defend the colony from the threats of a wide variety of predators, especially other ants. According to the season and the nutritional status of the colony, ants will periodically rear many reproductive individuals. These develop into winged males and females, which leave the colony in swarms. The winged ants mate and the fertilized females proceed to start new colonies. The story of ants' evolution in this world is a large volume of natural history punctuated by unbelievably intricate adaptations and equally unbelievable success. If the theme of that book is cooperation between individuals, its ink is the complicated chemical communication that makes the cooperation possible. Most ants are constantly emitting chemicals and responding to chemicals that carry more kinds of messages than we understand. Sound and sight have their roles among the ants, but they principally listen with their noses and speak with their glands.

A couple of those glands feed into a stinger in most ants. Many ants are too small to pierce human skin or are simply reluctant to sting a mammal. For the most part, we think of ants as household pests that annoy us during picnics or that infest our houses. Those common problems can be solved by keeping off the ground outdoors and by the sorts of structural modifications and insecticide applications discussed in chapters 4 and 5. In addition, baits can be very effective against ant colonies. Modern bait formulations use chemicals that are slow to act so that workers bring the food to the colony, killing many more ants than would a rapidly toxic active ingredient. People commonly make the mistake of using too little bait for the problem. They sometimes put out too few bait stations and then fail to replace them when the ants have taken all the contents. It's amazing how a huge household ant problem can be eliminated with the persistent use of commercial baits.

Although some of the household ants sting, it is much more common to encounter a painful stinging problem outdoors. Most of us have heard of "army ants," but this term refers more to the behavior of several groups of ants than to a particular kind of ant. They occur worldwide in the tropics as army ants, driver ants, safari ants, and numerous other local designations. They are defined by their nestless colonies and their group predation. Some, like the South American army ants, sting readily, whereas others, like the Asian and African driver ants, prefer to bite. In either case, it gives a person a helpless feeling to see a line of ants advancing, indicated not so much by the masses of ants, but by the leaping insects trying to escape these voracious predators. The best advice is to get out of the way.

The bullet ant of tropical Central and South America is supposed to have the most painful sting of any ant, bee, or wasp. They only live outdoors, in colonies of 100–2,000 big ants. Their neurotoxic venom causes not only significant pain, but also reversible systemic effects. These are large, conspicuous ants that should be easily avoided as long as there is light and you watch where you are putting your hands. Other ants in the same subfamily (Ponerinae) also have infamously powerful stings. The most characteristic native representative in the United States is in a group called the trap-jaw ants. These ants have huge jaws that they lock in place at a 180° angle, then slap shut; their jaws are considered to be the fastest-moving body part in the animal kingdom. The force exerted by the snapping jaws is enough to really skewer their prey. The American species, at least, can use their jaws to propel themselves a foot (30 cm) away from would-be attackers. An Asian ponerine ant, the Asian needle ant, was first noticed in the United States in Georgia in 1932, and has since become established from Alabama to Virginia in the Appalachian region. This small ant has a painful sting and is considered an emerging public health problem.

The harvester ants can cause real pain when they sting, even leading to systemic effects. They occur throughout the warmer regions of the United States down to temperate South America. The entrance hole to a harvester ant colony in the western United States is visible from a great distance because they carefully remove all vegetation and debris in a large circular area.

Fire ants usually present the greatest medical problems. There are native species in the United States and down to Costa Rica, but the two badly problematic species (red imported fire ants and black imported fire ants) came from South America less than a century ago. The red imported fire ant has since spread throughout the southeastern United States and the black imported fire ant is abundant in a small part of the south central part of the country. The commercial practice of renting bee colonies for almond pollination may have carried the red imported fire ant to California. California has mounted organized community campaigns to eradicate them, but the distribution of fire ants continues to expand there. They have also been introduced into southern Europe, southern Africa, Australia, New Zealand, Taiwan, and China. These ants live in large colonies that reach the surface as loose, earthen mounds up to one foot (30 cm) high. Fire ants vigorously defend their colonies following any disturbance, swarming over a foot, leg, or hand and inflicting many stings. Observed up close, the ants bite to gain the best leverage and then ram their stingers into the flesh. The toxins cause pain, followed by characteristic white lesions that resemble pimples. Anyone who lives near fire ants learns to recognize them and their mounds because most encounters are very painful. Allergies to fire ant venom are not uncommon, and a few deaths occur each year.

The United States and Taiwan have organized campaigns to reduce the number of fire ants or even to eliminate them. In addition to the local attempts to kill fire ants with baits in California, the U.S. Department of Agriculture organized the dissemination of pathogens that attack fire ants and of a small parasitic fly that eats fire ants. These biological control agents have spread throughout the southeastern United States. Hopefully, they will contribute toward the creation of an ecology similar to that of the fire ants' South American home, where they never reach the abundance they enjoy in North America. Taiwan retains a hope of eliminating the ants from its island nation through the systematic use of biological control agents and baits.

Highly effective baits are available to any homeowner, and you would certainly want to use them if your yard is infested. Fire ants are often attracted to electrical equipment, packing themselves into junction boxes and electric outlets. Treatment is difficult in those situations because the application of sprays can cause short-circuits. Dusts are safer, but if you are in doubt about the right treatment, call in a professional pest controller. If you are camping in fire ant country, you might want to spray down the ground around your sleeping area. Even a few stings by foraging worker fire ants are enough to disturb a good night's sleep.

BEES

Honey bees are one of the few insects that we actually cultivate agriculturally. They have been introduced around the world for their obvious benefits as honey producers and pollinators. As we all know, the workers can sting. They leave their stinger behind in the victim, carrying out an instinctive sacrifice of their own lives for the defense of the colony. The venom causes immediate pain, which from the standpoint of the bee is the whole point. Swelling and sometimes itching follow, making the whole experience unpleasantly memorable. Prompt removal of the stinger will limit the amount of venom pumped into your skin, and "prompt" here means that seconds count. A stinger removed within one or two seconds results in significantly less venom than one removed eight seconds after the sting. It doesn't matter whether you scrape the stinger off or pull it out with your fingers, the important thing is to get it out fast. For most people, the damage stops there, but some people develop an allergy to bee stings which can be life threatening. They should consult their physicians to see whether they should carry a kit with oral antihistamines and injectable norepinephrine, medicines that can reverse anaphylactic shock.

A single bee sting is no more than an inconvenience for most people. The problem is that the bee leaves behind a substance that alerts other bees to sting in the same place. If you are near a hive and the stinging starts, you could receive hundreds of stings and accumulate a lethal dose of the venom. The Africanized bees are particularly bad about this and may follow a person for many hundreds of yards before giving up the chase. With regular or Africanized honey bees, it is best to get as far away from the colony as quickly as possible once the bees decide to go on the defensive. Beekeepers have equipment for handling hives safely, but most of us do not.

Honey bees sometimes establish a new colony in an inconvenient location, like in your attic or in a tree in your backyard. if at all possible, it is the right thing to call your local beekeepers association to see whether someone will come and gather the colony. Domesticated European honey bees (unlike Africanized bees) are valuable insects, and it would be a shame to waste them. If no one is available to help or if you live in an Africanized bee area, you will have to call a professional pest controller to kill the colony. In the worst case, the hive may be tucked into a wall in your house. Between the removal of the wallboard to get at the colony and the damage from the accumulation of honey and wax, such infestations can be expensive to repair. The removal or control of a honey bee colony is the work of a professional. Never attempt this on your own.

Other bees can sting. Probably the most painful are the bumblebees, which most of us recognize as large, hairy bees with at least a bit of orange or yellow on them. They nest in the ground and are truly colonial, but the colonies are not nearly as large as those of the honey bees. Once disturbed, bumblebees can mount a significant defense of their nest. Occasionally, a person may be stung by a single bumblebee that is out foraging. The sting can be pretty painful and result in local swelling for a few days. Unless a person is allergic, the effects are minor.

Carpenter bees look a lot like bumblebees, except they are usually completely black and the abdomen includes areas that are completely bare. Males of some species look strange because they are hairy and big like a bumblebee, but they are completely pale in coloration. Few people welcome the females, not because they sting, but because they excavate their nests in wood. The holes usually enter the wood from the end grain, so that beams of garages and decks are common targets. The carpenter bees do not actually eat the wood, but the holes are large (about 0.5″ [1 cm] in diameter) and enough of them over years will ruin a piece of structural wood. A little bit of insecticidal dust or spray in the hole effectively kills the carpenter bees, though it is a shame to have to kill these interesting bugs. Painting or varnishing the wood will help to discourage other carpenter bees from burrowing in the same places.

The sweat bees vary in appearance from rather dull colors to pretty, metallic-colored insects that are attracted to the salt and moisture on your sweating skin. Sweat bees will sting in self-defense when they become trapped between your clothing and skin or when they feel threatened in some other way. The pain from these stings is very mild.

Stingless bees are sometimes raised for the small amount of honey they produce in small waxen cells. In South America, they are common natural inhabitants of crevices in walls or posts. They are really rather pleasant neighbors, always busy during the day, coming and going to the wax straws that form the entrances to their nests. When threatened, they attack by biting, causing minor irritation from secretions they produce in their mouthparts.

We hope that none of you are ever stung or bitten by a venomous arthropod. We also hope that you avoid the majority of blood-sucking bugs, especially those that carry disease-causing pathogens. Now that you have read about the world's current bug problems and their solutions, let's take a look at what the future may hold.

Box 13.1 Caterpillars and Moths: Dark Side of Psyche

Most people think of butterflies as the most beautiful of insects, moths as inevitable nighttime companions around lights, and caterpillars (the larvae of moths and butterflies) as occasional pests. This order of insects (Lepidoptera) has a few surprises: about 100 species worldwide have caterpillars or adults with irritating or toxic hairs; a variety of moths and butterflies feed on the wounds or eye secretions of mammals; and one genus actually sucks blood.

The most common problem is called *erucism* and is caused by caterpillars that possess urticating hairs or spines. The hairs vary a great deal in size and complexity, from short spines that only have a mechanical effect to large spines with venom reservoirs and pumping mechanisms. Usually, these urticating hairs are used in self-defense or as protection for some pupae. The most potent hairs cause extensive damage to the skin and a lot of pain. Occasionally, people develop allergies that cause even more serious problems. The smallest types of hairs are very abundant on the caterpillars, so abundant that, as the caterpillar changes skins, the hairs are released into the air and can cause widespread lung irritation. Some adult moths also have irritating hairs, but no venom, which causes an irritation called *lepidopterism*. Treatment includes removal of the spines using a fine tweezers or forceps. Like the smallest of cactus spines, small urticating hairs also can be removed by applying and removing adhesive tape to the affected area.

Most of the common caterpillars with urticating hairs are in one of three families: the slug caterpillars or cup moths (family Limacodidae, represented by the saddleback caterpillar in Figure 13.1.1), the giant silk moths (family Saturniidae, represented by the Io moth caterpillar in Figure 13.1.2), and the flannel moths (family Megalopygidae, represented by the puss caterpillar in Figure 13.1.3). The tussock moths (family Lymantriidae, represented by the browntail moth caterpillar in Figure 13.1.4) are also important, as are some species of tiger moths (family Arctiidae) and tent caterpillars (family Lasiocampidae). The only butterflies with urticating caterpillars are brush-footed butterflies (family Nymphalidae, represented by the mourning cloak butterfly caterpillar in Figure 13.1.5).

Figure 13.1.1. Saddleback caterpillar (Limacodidae: *Acharia stimulea*; up to 1″ or 2.5 cm long). Photo by Lynette Schimming.

Figure 13.1.2. Io moth caterpillar (Saturniidae: *Automeris io*; up to 2.5″ or 6.4 cm long). Photo by M. C. Thomas.

Although the saddleback, or packsaddle, caterpillar is native to eastern North America, it is found worldwide. The saddleback caterpillars are primarily green and easily identified by the brown or purplish saddle-like mark on their backs. They have multiple stout, venomous spines along the sides of their bodies. Their stings are painful and can cause swelling and nausea and leave a rash that lasts for days. Other species of slug caterpillars occur elsewhere in the world, including some in Australia that leave large flat wheals and reddened skin.

The Io moth caterpillar is a spiny green caterpillar with a white broad line along each side, bordered above and below by thin

Figure 13.1.3. Puss caterpillar (Megalopygidae: *Megalopyge opercularis*; up to 2.8″ or 7 cm long). Photo by Sturgis McKeever, Georgia Southern University, Bugwood.org.

Figure 13.1.4. Browntail moth caterpillar (Lymantriidae: *Euproctis chrysorrhoea*; up to 1.5″ or 3.8 cm long). Photo from www.winchester.gov.uk/NewsArticleR.asp?id=SX9452-A77F6768.

Figure 13.1.5. Mourning cloak butterfly caterpillar (Nymphalidae: *Nymphalis antiopa*; up to 1.5″ or 3.8 cm long). Photo by Charles Lewallen.

reddish stripes. It also has a row of tubercles armed with spines that are connected to venom glands. These spines are toxic and can cause severe pain when they get in contact with human skin.

The puss caterpillar is teardrop-shaped and heavily covered with long, soft hairs that resemble a tuft of cotton or fur. Beneath the hairs are many poisonous spines that cause severe pain and rash. The puss caterpillars can sometimes cause cardiovascular and neurologic symptoms. In areas with many cast skins, the spines can become airborne and cause lung irritation in many thousands of people. A species in the same family in South America, sometimes called the *tatarana* or *yso tata* (literally, fire worm), causes intense pain on contact.

The browntail moth caterpillar is native to Europe and was introduced to North America in 1897. Its hairs can cause an irritating skin rash on humans similar to that caused by poison ivy. The mistletoe browntail moth is a similar species in Australia.

The best protection from urticating caterpillars is to avoid them. Education and an awareness of the hazards are usually sufficient to prevent repeat exposures. However, when outdoors in areas where these caterpillars are present, wear gloves, a long-sleeved shirt, long pants, socks, and boots.

The blood-sucking moths are certainly one of the oddest corners of entomology. Restricted to Southeast Asia, they have the typical coiled beak of their kind, but with one exception: the tip is pointed and barbed to penetrate thick skin. They seem to represent a line of evolution that progresses from feeding on animal eye secretions (practiced by many moths and butterflies), to feeding on wounds, to actually piercing the skin for blood. Reports from Australia and from Finland suggest that the blood-feeding habit there is more widespread among moths and butterflies. We hope that more are not on the way.

Psyche means "butterfly" in Greek and was the name of the enchanting mythological character who was Cupid's immortalized mate (and mother of Bliss). *Psyche* is also the title of the Cambridge Entomological Club's journal, which was established at Harvard College in 1874, when people really appreciated the classics.

14 FUTURE DIRECTIONS

We have given you the tools you need to avoid problems with blood-sucking and venomous arthropods. Most of those bugs will not change in the future and neither will the damage they do to people. On the other hand, the extent of each kind of problem is in constant flux as we find better ways to deal with it or as we make the problem worse through our own activities. There are many famous examples of these kinds of changes in human history, including the conquest of epidemic typhus with insecticides and antibiotics, the control of yellow fever with vaccines and mosquito control, the expansion of dengue throughout the tropics since the 1980s, and the resurgence of malaria in Korea since 1993. Fortunately, science, government, and the engine of the marketplace have driven the development of many new tools since the 1940s. It is hard to imagine a world without aerosol cans of insecticide, long-lasting repellents, and a variety of insecticides for home use, yet all of those inventions came along in the last 70 years. What can we hope for in the next 70 years?

New insecticides are incredibly expensive to develop because of the costs of finding them and the costs of showing that they will not harm the user or the environment. Cost estimates depend on the kind of chemical and the kind of company doing the work. Ten million dollars is a low figure and $300 million is a high one. It is perfectly understandable that companies think carefully before embarking on the development of a new product. They want to be sure that the product will sell, that it will not produce any damaging surprises, and that no one else is going to compete by producing the same thing. Market forces drive new invention, and business caution appropriately restricts new development. Some insecticides, like allethrin, have been around for decades; however, we would find that we have fewer and fewer choices if we never developed new ones. Two of the major factors that take insecticides off the market are new legitimate safety concerns and newly developed resistance by target pests.

There is certainly room to make development more efficient. International cooperation, enlightened regulation, and the good intentions of nongovernmental organizations are all forces actively driving attempts to share data, harmonize requirements, and target the needs of disadvantaged populations. The temptation for industry is to develop insecticides for use against a wide variety of insects, spreading the investment across many pests. This strategy tends to produce chemical hammers to which every pest is a nail. The trend of more thoughtful research is to produce new chemicals that can be targeted at specific pests, either through formulation or mode of action. A wider variety of insecticides complicates recommendations, but the greater choice would give the public a tool box that has a hammer for a nail, a screwdriver for a screw, and a wrench for a bolt. By using the right insecticide for a precise purpose, its application is likely to be more economical, safer, and more effective.

The development of skin-applied repellents has really accelerated since the late 1990s. Even though new active ingredients have to compete with some very effective compounds already on the market, promising new plant-derived compounds are under consideration. Formulation has gotten more sophisticated with the introduction of liposomes that surround the active ingredient in a way that improves both the repellent's duration and its resistance to absorption. Special formulations and new active ingredients add to the cost of a product, but they may also expand the current market in repellents. Presumably, people would use repellents more often if they could depend on improved performance and application characteristics appropriate to their need. Scientists in the laboratory are rediscovering the power of mixtures to enhance the repellency of active ingredients, particularly plant-derived compounds. Some of the preliminary results suggest that an excellent, long-duration repellent might contain as little as 5% active ingredient. Such a product would open the door to many new kinds of formulations that simply will not work when a product contains 30% active ingredient. We could see effective repellents applied like a convenient deodorant or cologne.

New spatial repellents, the kind that keep biting pests out of a larger area, are the objective of much research. Mosquito coils, citronella candles, and granular products are old, familiar examples. The public really likes these kinds of products, even though they are incompletely effective. Some of the newer active ingredients, like certain forms of linalool and volatile pyrethroids, seem to do a better job. The incremental improvement since the 1990s suggests that we may get to the point where the cute little solar-powered device, wristband, candle,

Box 14.1 Cheap, Effective, Easy to Use: A Repellent for Those Who Need It

At this writing in 2008, a very low-cost repellent lotion (16% PMD and 5% lemongrass oil) has been developed for global distribution to malarial countries. The inventor of this product is Sam Darling, the grandson of one of the pioneers of tropical medicine in the Americas, Samuel T. Darling. Inspired by his grandfather's commitment to public health and his efforts to combat malaria among the poor, Sam directed a research effort that developed a repellent with a minimum of active ingredients and a maximum of protection time. By doing this, he made his repellent affordable for poor, malaria-endemic communities. His repellent has demonstrated complete protection for nine hours against the main vector of malaria in Africa. It has performed equally well against the vectors of dengue and West Nile fever viruses.

The new formulation has a pleasing odor, is not oily, and is easy to apply. With an average cost (including delivery to Africa) of 2.6 cents per person per day, this product offers the promise of a significant reduction of malaria when used with insecticide-treated bed nets. It could become part of an integrated program to eliminate malaria from an area if it were combined with other measures, like indoor residual insecticides and community larval control. Even by itself, the repellent could improve the health of those most at risk from malaria. Perhaps good intentions will triumph among all the government and business players who will be necessary to make Sam Darling's product work. Time will tell whether a good product can be combined with an altruistic business model to protect the people who most need the help.

or treated clothing actually will provide something close to complete protection.

Traps also attempt to provide protection over a large area. None of them are simple to use effectively, though they are useful when applied in the right situation and when there are enough of them. It is hard to imagine a trap so powerful that a single device would divert all the biting flies, fleas, or ticks from their intended hosts. Perhaps the more likely improvement is the establishment of firm guidelines for the most effective use of traps in conjunction with other control measures. The technology at least partially exists for a mechanical "buddy" that would take the place of those unfortunately attractive individuals who always seem to get the bites.

The management of blood-sucking and venomous pests will be improved by the invention of new chemicals and devices, but information-based improvements are likely to have much more dramatic results. Intelligent, logical, targeted use of the existing tools would probably solve most of our pest problems. Scientific development of the relevant information about risk assessment, surveillance, control, and sustainability is only part of the effort. The information must be organized and communicated so that professionals at all levels can find ways to apply their skills in a coordinated way for maximum public benefit. The list of those who need to cooperate ranges from physicians to pest controllers and from government regulators to manufacturers. However, the list does not end there; it ends with the individual. *You* are the person most responsible for avoiding arthropod-borne diseases, blood-sucking pests, and venomous bugs. Arming yourself with information and exerting influence on your community are the keys to the best protection in the future.

PUTTING IT ALL TOGETHER

Let's meet the Carters. They are a fictional American family with real bug problems. They solve these problems by identifying the pest, evaluating the problem, eliminating sources, and making intelligent use of insecticides, clothing, and repellents. George and Melissa are the grandparents in their vigorous early 60s. George teaches at the state university and Melissa runs a local business. George and Melissa moved into a condo on an upper floor close to the middle of San José, California, when the kids well and truly left the nest. Their three children live all over the place.

Their daughter, Cary, is a 36-year-old mother married to John Monroe, a broker in San Antonio, Texas. It's a busy household with 10-year-old Francy, 6-year-old Doug, and one more due in about seven months. They all live in a small 1920s bungalow built on cedar posts in a central neighborhood. Cary and John bought the house shortly after they got married. It was all they could afford at the time, but they have really fixed it up into a convenient little palace.

Shawn is 29 years old and a newly promoted sergeant in the U.S. Army. He's been in Iraq for 12 months and has recently taken charge of an eight-man squad. Their main job is the grunt work of patrolling and manning checkpoints, but they do their share of camp maintenance and "other duties as assigned." The work is scary and boring by turns. Although they usually return to relatively comfortable quarters on base, they often end up sleeping outdoors because it is either too hot to continue or too dangerous to return at night. Shawn is a good soldier who pays attention to his surroundings and his health. He is just beginning to learn that he has to do the same for his men.

Frank, 26, got a master's degree in business and then surprised everyone by joining the Peace Corps. The organization sent him to Paraguay, where he is just finishing his two-year stint. He's gotten pretty good at working with local officials to identify profitable markets. Although he has traveled all over the South American country for his work, he is based in Atyra, a rural center with perhaps 10,000 people, 2,000 pigs, 1,000 cattle, 75 donkeys, and a population of chickens that fluctuates wildly according to the proximity of a holiday.

George and Melissa love to go places outdoors. Ironically, one of the reasons they gave up their suburban house was to spend more time outdoors. They especially like to go hiking and camping in their own area, which has habitats ranging from desert-like conditions to redwood forests. The two of them hardly think about pest problems at home now. George doesn't miss the days of cleaning out the black widows from the corners of the garage or from the crevices in the wooden fence, always afraid he might miss the one spider that would transform one of his children into a national, although exceedingly rare, statistic. Spiders sometimes got in the house, too. Recognizing a black widow was easy, and he had grown up learning that they were dangerous. It was only when the kids were a little older that he learned about the yellow sac spider. He only had a vague picture on a brochure to guide him, but it was enough to keep him from trying to kill every household spider in fear that one of them might be a problem. Once in a while, he killed a light-colored spider with longish, marked legs, but he left the others alone in their corner webs. Now, he has heard that the yellow sac spider and the hobo spider to the north are not such problems after all—two more examples of applied science changing its advice to the public as new studies are performed. On the whole, George liked spiders when he had his house. There always seemed to be a fly or two around, but he could see that most of the spiders were busily keeping the numbers down. At least in San José, he doesn't have to worry about the brown recluse.

Mosquitoes are not a bad problem for George and Melissa, and they never have been. Some people used to think that the dry summers kept populations down, but many mosquitoes actually thrived as the streams turned into a series of still pools. Three species of mosquitoes seemed to specialize in cool-weather development, taking advantage of the winter rains. One in particular was a plague that could emerge in huge numbers from the small patches of impounded salt marsh along the San Francisco Bay. The solution to these potentially bad mosquito problems was a mosquito abatement district that had been in operation for over 70 years. George saw the cost on his property tax bill every year: $23. The cost was so low that he hardly paid attention, and it was far less than the money he would have spent on repellents and home spraying. The mosquito abatement district had its greatest effect by finding larvae and then killing them with environmentally friendly Bti, a bacterial toxin that only kills mosquitoes and related flies. A few years back, when West Nile virus came roaring through the bird population, the abatement district did some application of insecticidal fog at night in areas where it had found infectious mosquitoes. This procedure seemed to work, since there were hundreds of infected birds but only a few human cases. Nonetheless, Melissa swore that they would get the West Nile virus vaccine when it came out.

These days, George and Melissa mainly worry about tick bites. Often hiking where deer are abundant, they realize that California has its own species of tick that transmits the Lyme disease bacteria. They also know that a different kind of tick transmits the bacteria that cause spotted fever and ehrlichiosis. Their area has few cases of these diseases, however, and the actual risk is really quite low. They take the precaution of wearing long trousers that they treat with a permethrin spray. Usually, they also spray the tops of their boots and socks with an aerosol repellent. When they get home or back to a campsite, they look each other over for ticks, but the clothing treatment pretty much eliminates the problem. One time, they found a tick wandering around in their bedroom, probably carried back on a jacket or pack that had rested on the ground. Melissa washed all the hiking clothes and dried them on a high heat, which certainly would have killed any other ticks.

Down in San Antonio, Cary's family does not have it so easy. There is a good reason that pest control companies take up a large proportion of the phone book's yellow pages. When Cary and John first moved into their house, there had been problem after problem. The previous owners had evidently tolerated an arthropod zoo on their property, feeding many of the animals with their own blood. First, the mosquitoes were annoying much of the year, some biting during the day and some biting at night. One little dark one with white stripes even followed them into the house, laying its eggs in the rough calcification near the rim of a glass they used to rinse their teeth. The trash pile at the back of the lot encouraged all kinds of bugs and seemed to be the source of roof rats that went in and out of the numerous small openings in their old house. One hot night shortly after moving in, Cary made the mistake of walking onto the front lawn to see the stars. She stepped right on a fire ant mound, the inhabitants of which objected to the damage with hundreds of stings that became solid patches of festering sores. As if that weren't enough, John noticed a scorpion in their cockroach-infested kitchen. He went into the attic to see if they were coming from there. Sure enough, there were several in flashlight range.

Cary and John were not the kind of people to panic, but they knew they had to figure out how to make their house more livable for people and less livable for pests. They performed most of the work themselves because they didn't have much money. John talked to neighbors and to salespeople at the big-box home store, but he never felt really sure that the advice was reliable. All were well-meaning, but the neighbors seemed a little too willing to wait until a problem was completely intolerable before taking action. The salespeople's knowledge was based on what they heard from other customers and from what they read on labels; sometimes they were right and sometimes they were wrong, but they always had a product that they said would be worth buying. The trips to the store were helpful, though, because John picked up a few brochures on pest problems. These listed a county office as a source of advice. Cary gave them a call and was surprised to receive a torrent of questions about their pest problems. Fortunately, she knew one bug from another, and the pest management expert soon realized he could get sensible answers to his questions about size, color, and shape. Together, they worked up a plan that solved the easiest problems first. The solutions stressed permanent structural changes to the property, aimed at reducing the expense of repeated pesticide applications.

The fire ants were easily discouraged with a couple of spoonfuls of toxic bait on each mound, placed there in the morning after the dew had dried from the grass. After the first treatment, it was just a matter of keeping up with the mounds as they appeared. The general clean-up of the property was a necessity just to make the place decent—and it was a lot of work. Cary and John were surprised to see hordes of mosquito wigglers in old plastic buckets and discarded cups that contained small amounts of water provided by the neighbors' sprinklers. They also saw rats scuttle off, abundant spiders, and scorpions. At that point, they started wearing boots and leather gloves to do the picking up in the yard and under the house. Inside the house, the couple undertook a strict regimen of cleaning the kitchen and eliminating any other source of cockroach food every night. They put a new washer in the dripping kitchen tap to deny the bugs an easy source of moisture. Cary bought a couple of packages of cockroach bait stations and put about 12 of them at the corners of the walls and the floor, in cabinets, and at the very back of counters. While she was crawling around with the bait stations, she noticed a nest of cockroaches on the underside of the small kitchen table. Cary took the table outside, sprayed the underside with a phenothrin-based product (labeled for use around food), and left it out in the hot Texas sun for a while. With the cockroaches gone, the scorpions quit wandering into the kitchen, and the place was much cleaner. John carefully applied a pyrethroid-impregnated silica dust to the floor of the attic to eliminate the area as a source of scorpions in the future.

Rats and mosquitoes were more persistent problems. The pest management advisor suggested doing the tedious business of sealing every entryway into the house. When they looked carefully, Cary and John were surprised to see gaps around every pipe, wire, and vent that entered the house through the walls or the floors. It took some creativity, but they finally managed to seal all the gaps smaller than a dime with caulk and the larger ones with a combination of cement and steel wool. Some of the screens covering vents to the attic had gaping holes, which they fixed with stout hardware cloth. The window screens did not seal tightly to the window frames, and many of the screens had holes, leading to temporary fixes with tape and a long (monetary) campaign of replacement. Making the house rat- and mosquito-proof took a while, but eventually it was done. The mosquito problem indoors got better right away, though they still heard rats scuffling around at night. Cary and John put traps wherever they thought the rats were active. This led to a few exciting days punctuated by the periodic *snap!* of a trap. It really gave the couple a sense of accomplishment to evict these destructive tenants. It was with some pride that John took a picture of Cary holding up a trap with a dead rat, its head crushed by the device. They were always careful to put the rat in a plastic sack before throwing it in the trash and never touched the animal directly. A bloody trap went into a bucket with chlorine bleach solution to disinfect it, and they washed their hands thoroughly afterward. Fortunately, they did not have to use poison rat bait because they had done a good job of sealing the house and of eliminating hiding places outside. Although disposal of the dead rats in the traps was not pleasant, that was a lot better than having dead, poisoned rats stuck in the walls and raising a stink (and possibly flies) for a couple of weeks.

At this point, the fire ants, scorpions, and cockroaches were gone. The mosquitoes only bit outdoors, which was bad enough, but at least you could work outside by putting on repellent. If they were going to be out for a while working on the garden, they used a formulation that had a bit of a smell and cost a lot per bottle, but it lasted pretty well through hours of sweaty labor. Cary kept an aerosol can of a product that had very little odor on the not-yet-screened porch. Sometimes, they would sit out there for a half hour after sunset, fending off the mosquitoes with a light repellent spray that they hardly noticed on their skin for the rest of the evening before showering and going to bed. The rats were out of the house and visited their tidy yard much less often, but now a new problem emerged. Cary was working in her home office one day and noticed a distinct, irritating bite on her arm. She kept getting these bites for a few days. The welts persisted almost a week as raised, red bumps—and they itched. Cary called her trusty pest management professional, who never felt like he was doing his job unless he answered a question with a question. He asked whether she had seen any rats lately. When she said that she had not, he explained that rat mites were using her as a blood source now that their rodent hosts were gone. However, rat mites can't survive on human blood meals, making the problem a temporary one. In the meantime, she could get some relief by spraying the affected room with an indoor insecticide and by using a repellent.

Cary and John continue to clean, landscape, and fix their house. They find a wasp nest here and a black widow there, but they now have their property to the point that they are not producing their own pests.

Shawn, on the other hand, is not having an easy time in Iraq. The tension, boredom, and heat are bad enough, but he and his squad have other problems as well. When they are at the base, things aren't too bad because contractors and preventive medicine units work on the bug problems. On patrol, however, they sometimes encounter nasty biting pests. Early in the deployment, one soldier was digging a position into the loose rock and dusty soil at the outskirts of Baghdad. He was hot, his helmet was heavy, and his body armor felt like a wool blanket on his sweating body. It wasn't easy to do the task, let alone be careful of where he put his hands. Moving a rock, he got the briefest glimpse of a brown crab-like creature before he felt a sharp, searing pain on the back of his hand. The soldier shouted involuntarily, and Shawn rushed over to see the problem. He had listened intently during the briefing on scorpions and been fascinated by the poster on venomous spiders and scorpions that was posted in the clinic. Shawn managed to kill the scorpion with his rifle butt, providing a specimen for identification later on. More important, he had the good sense to call in a medevac helicopter to evacuate the soldier right away. The rest of the squad was a lot more careful after that. They were impressed by the horrible black wound that stretched inches along the hand and arm of their comrade in the field hospital. Unfortunately, the wound got worse over the next few days, and the decision was made to send the poor guy to a major medical facility in Europe. They haven't heard whether or not he needed skin grafts.

Following the rains in late winter, there are week-long outbreaks of mosquitoes that look like they are dressed in tan camouflage. A little repellent is enough to discourage the worst of their bites, but most soldiers do not bother. Wearing long sleeves and long trousers during work and the use of bed nets during sleep are enough protection against a temporary mosquito problem with no risk of disease.

The sand flies are a much more serious problem. Despite attempts to suppress the pests by applying insecticidal fogs to the camp and permethrin on tents, the small biting flies continue to plague soldiers who sleep anywhere except a tightly sealed, air-conditioned structure. Repellents keep them off, but it is pretty miserable to use the toothpaste-thick, smelly military repellent every night—even though the stuff continues to provide protection all night. A lot of people get ugly sores from the parasite carried by the sand flies, requiring extensive treatment with a harsh drug.

Melissa sent Shawn a wide variety of repellents when he asked for them. A cute device that heats a strip of paper with active ingredient is neat, but it can only be used outdoors and it requires a big stack of treated strips to be used all the time. He has found that 20% IR3535 spray or a special new formulation of DEET are the easiest to use and most pleasant to wear, and they prevent bites for the whole night. He also uses a military-issue indoor aerosol insecticide before going to bed. Biting pests aren't the worst thing about being in Iraq, but they sure are a little more sour icing on the bitter cake. George and Melissa are very proud of their son and his service, but Shawn sure is looking forward to the end of his tour in a few months.

Frank and Shawn have always been close and they e-mail each other regularly, or as regularly as Frank's meager wages allow him access to the single Internet café in Atyra. Frank worries about Shawn for obvious reasons, but Shawn worries just as much about Frank. After all, Frank doesn't have the U.S. Army looking out for him. When Frank describes his bug problems in Paraguay, Shawn is really impressed. Frank is living with nature in a way that most Americans do not understand.

His small house has crumbling brick walls, no glass windows or screens (just wooden shutters), a door with a 1″ gap below, and openings between the uninsulated tile roof and the walls. Frank might as well be sleeping outside with just a roof over him. He does his best to protect himself, and in the event, that is enough. First, he takes the malaria pills that the Peace Corps gives him, in spite of sometimes disturbing side effects. He sleeps in a bed that supports a thin mattress on a springy metal screen, putting a foot of space between his body and whatever is crawling on the brick floor. He wears shoes and uses a flashlight whenever he has to walk across the yard at night to use the outhouse at the other end of the property, a behavior reinforced by regular sightings of scorpions on the walkway. Frank sleeps every night under a bed net to reduce the possibility of kissing bug bites. In fact, he carries his trusty bed net with him whenever he travels. He knows that kissing bugs are around because he often sees them in his neighbors' houses, hiding in the cracks in the walls, behind furniture, and under picture frames. It is a shame that the bugs are so common because the blood-sucking pests are the essential link in maintaining the high prevalence of the incurable Chagas disease. Frank got the Peace Corps to pay for the plaster he applied to the inside and outside of his house. The elimination of cracks in the walls cut down on the number of kissing bugs, and he has been pleased to see that some of his neighbors have followed his example by plastering their houses.

Melissa also sent repellents to Frank, which was a great luxury in a country where the reliable Brazilian products are quite pricey. He likes pump sprays that he can apply to the lower part of his trousers and his socks while he sits at his favorite bar-restaurant enjoying a chicken leg, *mandioca*, and a beer. Otherwise, he would come home with welts from the yellow fever mosquitoes, which like to bite under the table. Periodically, towns in Paraguay suffer outbreaks of dengue fever, giving Frank an even better reason to avoid bites from this mosquito. Some of the local people burn mosquito coils to keep the mosquitoes off while they eat or watch television in the evening, but the smoke makes Frank choke, and he prefers the repellents his mom sent.

Although Frank sees a lot of other biting and stinging arthropods, most are not a problem every day. A couple of kinds of deer flies attack passersby down by the stream. Paraguayan deer flies do not have the persistence of their kind in other parts of the world, and they are easily brushed away either before or after they nip the skin. One day out in a field, Frank felt privileged to witness a sheet of army ants coming toward him. He was able to step out of the way, unlike the grasshoppers and crickets that were fleeing for their lives. Frank knows that Africanized bees are the rule rather than the exception in his area, though they do not seem to be any special hazard. He sees them on flowers like any other bee and a local beekeeper has mastered methods of handling the insects when they are grouchy. Kids in the village, who spend a lot more time outside, are careful to avoid the frequent swarms of bees. If one gets stung, the whole gaggle of children run away as fast as they can before other bees join the defense.

Frank also has had personal experience with a human bot fly growing in his arm and with chigoe fleas growing under his toenails. Fortunately, both were rare, though uncomfortable events. His neighbor showed him how to suffocate the bot fly larva with tape, making it easy to squeeze it out of its chamber. The chigoe fleas were trickier because their chambers tend to get infected, but with rubbing alcohol and a sterile needle he was able to enlarge the hole a little and neatly hook the expanding flea out of the toe.

George and Melissa greatly respect their sons' adventures. They got to thinking that they should combine their own interests in the outdoors with a trip in the wild. They were in a mood to do it up right, so they planned a combination of cultural sites in the Mediterranean region with a safari in Kenya. George had the generous idea of inviting their daughter, Cary, and their eldest grandchild, Francy, on the "trip of a lifetime." Cary said she would not go because (a) she had work to do, and (b) she did not like the idea of exposing her body to the various bites, stings, and attendant diseases of the region during her pregnancy. As soon as Francy heard the idea, though, she was wild to go. Eventually a compromise was reached, and Francy flew back from Cairo prior to the part of the trip in Kenya, where there would be a much larger risk of exposure to diseases, including malaria, tick typhus, and sleeping sickness. Melissa had made sure that they took repellents and aerosol insecticide with them, as well as the antimalarials and vaccines recommended by their physician. In retrospect, Melissa felt that her preparations were well justified even in southern Europe when mosquitoes became an annoying problem. The repellents helped in Egypt as well, though one outing was nearly ruined by the hordes of biting midges, which seemingly emerged from the sand.

George and Melissa reluctantly put Francy on a plane in Cairo after two weeks of traveling in Italy, Turkey, and Egypt. They had taken pains to do things a 10-year-old would enjoy, and Francy was only a little homesick. The adventure part of their trip is about to begin. As soon as they land in Nairobi, they know they have departed from their normal experience and, to some extent, from safety. Hundreds of thousands of tourists enjoy Kenya every year, but the travel warnings and sparser infrastructure give the place an edgy feel. They take a cab to the middle of town, arriving in the cabin-like room of a charming hotel. Everyone is more than friendly and helpful, including the porter, who carefully places an insecticide-treated strip on a permanent electric fixture in the wall, where it will emit fumes all night as protection from the mosquitoes. George is suspicious of breathing he-doesn't-know-what all night, so he removes the strip, sprays some aerosol insecticide in the room, and hangs their new conical bed net from a light fixture. It is probably overly cautious in this high-elevation city, but he figures he might as well start as he means to go on.

Their first stop is a resort on the Indian Ocean near Mombasa. The place is almost too beautiful, with shining white sands, flowering tropical trees, neat white buildings, and an impossibly blue sea. It is really set up for the guests' comfort. A spacious bedroom with a tile floor leads onto a balcony overlooking the beach. The room has an electric fan, but no air conditioning, and it really doesn't need it. A gentle breeze comes through the open door of the balcony. A mosquito net is carefully tucked up above the bed and a mosquito coil rests in a sparkling clean ceramic container on the glass table on the balcony. It is clear that mosquitoes are part of the experience in paradise. George and Melissa enjoy a few days at this resort, swimming, boating, and taking a tour through local villages and forests. They get a few mosquito bites during the day, but they avoid most of them by spraying their arms and legs with repellent. Rather than waiting until they notice the first bites, they carefully reapply at the intervals suggested by the label, minimizing the chance of getting dengue virus. At night, they are able to sit out on the balcony with their drinks as the coil does its work in the still air. The bed net is great protection against the mosquitoes that they can hear buzzing against the ceiling.

They are really ready for the final part of their vacation when the four-wheel-drive Land Rover comes to pick them up for a week-long safari. The trip to the Rift Valley takes the better part of the day, but does leave them time to settle into one of the huts. They are part of a group of nine tourists whose goal is to see as many of the large mammals and interesting plants of the area as possible. Guided by experienced people and blessed with good weather, they have a good chance of succeeding. During the day, they are carted around in an open truck to places that are safe and that offer the best chance of seeing the widest variety of wildlife. They return to the camp during the heat of the day, get dinner, and then head out again to elevated platforms that are secure from the large predators. It is a rigorous routine, relieved by simple delicious food and deep exhausted sleep.

The huts are well equipped with mosquito coils and bed nets. George and Melissa have already read up about it, but some of their fellow travelers are surprised to learn from the guides that the area is ground zero for at least four serious diseases associated with arthropod bites: tick typhus from hard ticks, leishmaniasis from sand flies, sleeping sickness from tsetse flies, and malaria from mosquitoes. Melissa and George are well prepared. They always wear permethrin-treated long trousers, use the longest-lasting repellent regularly on exposed skin, and sleep under the fine-mesh bed net they brought along. Before going to bed, they spray the hut thoroughly with an aerosol insecticide, and they use a little repellent on their knees and elbows, which might contact the net during the night. They are being extremely careful, hardly noticing any bites despite the abundant mosquitoes and sand flies they encounter. Some of the other travelers have not been as fortunate. One man decided that the bed net was too confining and ended up covered in welts the next morning. Another couple is relying on a locally produced botanical repellent that provides little protection. Some of the people insist on wearing shorts in the field. They get more sunburn and have to pull off ticks every evening. The guides take them to only one place with tsetse flies, a streambed frequented by many animals. It is a nice shady place with a good look at local trees that only grow near water. George and Melissa are surprised to find that their repellent only keeps the tsetse off for about an hour. Out come the long-sleeved shirts, which protect them enough to make the visit a pleasure instead of a penance.

The experience in the bush is delightful. In one short week, they come to appreciate the value of a truly wild habitat, and they start to see the world from a viewpoint outside the human perspective. George and Melissa are pleased that they have suffered no fever, rash, or sore in the weeks following the trip. Quite rightly, they figure that their precautions worked. Their luggage was not so lucky. As soon as Melissa saw a bed bug while she was unpacking, she closed everything up, put it all in plastic bags sealed tight, and took them outside. She slipped a no-pest strip in each bag and left them outside the condo in their storage locker for weeks. When Melissa finally took out the luggage and aired it out, she was pretty nostalgic about all the experiences they had enjoyed. George and she started thinking about their next adventure trip, possibly to Northern Queensland, Australia, where they will be able to exercise their precautions against dengue-carrying mosquitoes in Cairns, scrub typhus chiggers in the pristine rainforest, and the painful bites of horse flies at Crater Lakes National Park.

The group naturally had bonded together and kept in touch for a while after the safari. The man who received all the bites on the first night got to stay in Kenya for an extra week. On the day he was supposed to get on a plane headed home, he suddenly felt extremely ill with high fever, splitting headache, and nausea. Desperate, he went to a clinic in Nairobi, where the doctor immediately identified malaria as the cause. Blood samples confirmed the presence of the parasites, and he was

Box 15.1 **Bloodsucking and venomous pests present complicated problems for people. You can go a long way toward solving these problems by approaching each situation systematically, following detailed advice presented throughout this book. Use the flow chart below to get an idea of where to start and how to proceed.**

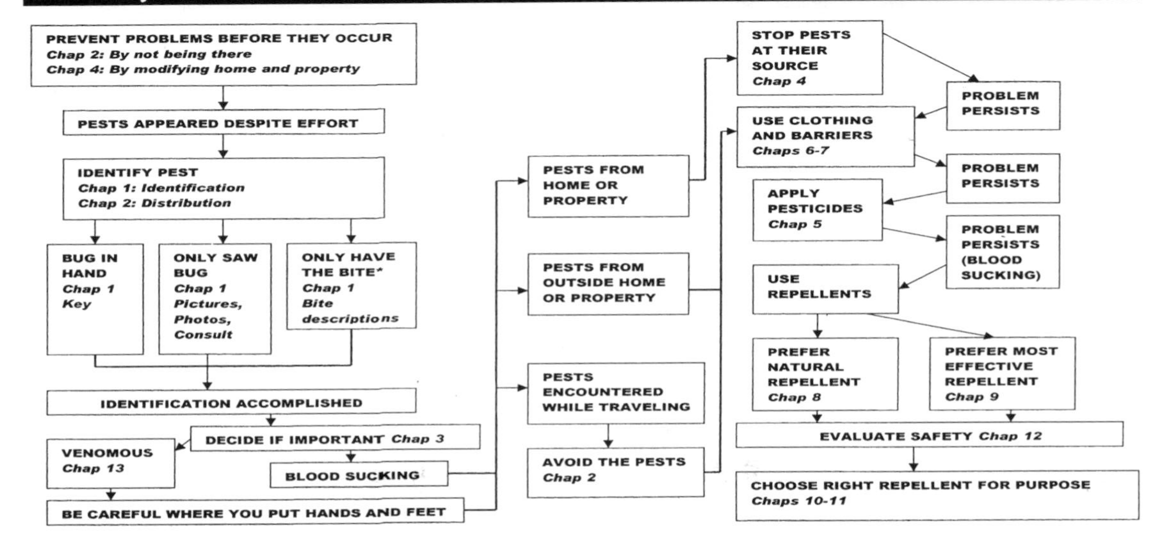

*Seek medical help promptly if the bite appears infected, badly swollen, persistently painful, progressively worse, or otherwise suggests a serious condition.

quickly hospitalized and treated. It was an experience he and his finances will not forget, and one he could have avoided.

Shawn and Frank are finally headed home. George, Melissa, Cary, and John are thrilled that the two brothers have come through their adventures unscathed. They will have a grand homecoming at Cary's house in San Antonio, which is close to Shawn's army base and with a little more space than the condo in San José. Besides, Cary is not anxious to travel when she is this pregnant. Everyone is so thrilled to see each other that they don't notice at first how introspective Shawn has become and how Frank is bursting to reveal a secret. They exchange stories all evening about their foreign experiences, complete with scorpions, spiders, and mosquitoes. Cary remarks that, living in San Antonio, the pests don't sound all that foreign to her. After the final bite of dessert, Frank reveals that he has an announcement. He will be returning to Paraguay to marry a Paraguaya and take up a post with the embassy. Francy jumps up and states that she has an announcement too. All eyes turn to her as she proudly proclaims, "I'm going to grow up to be an entomologist!"

GLOSSARY

active ingredient: a component in an insecticide or repellent formulation that provides insecticidal or repellent activity against target organisms.

allergen, allergenic (adj.): a substance that can produce a hypersensitive—but not necessarily harmful—reaction in the body. Some common allergens are pollen, animal dander, house dust, feathers, some insect hairs and scales, and some foods. The body normally protects against allergens using the immune system.

anaphylactic reaction: a serious hypersensitive condition induced by contact with certain antigens.

antibody: an immunoglobulin produced by lymphocytes in response to bacteria, viruses, or other antigenic substances.

antigen: a substance, usually a protein, which causes the formation of an antibody and reacts specifically with that antibody.

arthropod: a member of the phylum Arthropoda, which includes crabs, lobsters, mites, ticks, insects, and spiders. They generally have a jointed exoskeleton (cuticle) and paired, segmented legs.

Bti: abbreviation for the bacterium *Bacillus thuringiensis israelensis*, an active ingredient used in insecticides to kill mosquito larvae.

cuticle: the exterior covering of an arthropod, usually leathery or hard and serving the functions of both skin and skeleton; hence the alternative name, exoskeleton.

ecological: of or pertaining to ecology; conditioned by the biotic surroundings.

ecology: the physiology of organisms in relation to their environment; the reaction of organisms to the conditions of their existence, including the modifications of these reactions in relation to changes in the environment.

entomology: technically, the branch of zoology that deals with insects only; in practice, often applied to the study of all arthropods, including the application of the science for human benefit.

EPA: abbreviation of U.S. Environmental Protection Agency.

eschar: a scab or dry crust on the skin resulting from a burn, infection, or skin disease.

evolution: the theory of the origin and propagation of all biological species and their development through natural selection of variants over long periods of time.

immune system: a biochemical complex that protects the body against pathogenic organisms and other foreign bodies. The human immune system responds to many kinds of foreign substances in specific ways, using a combination of soluble proteins (antibodies) that stick to the substances and specially primed cells (among them, some of the white blood cells). A primary immune response may be general and not based on experience with that particular substance. Subsequent exposure trains the immune system to respond more specifically, sometimes producing an allergic response or other symptoms that cause illness. In many cases, repeated exposure over a long period adjusts the immune response to a non-allergic state.

immunity: the quality of being not susceptible or being unaffected by a particular disease or condition.

infected: the condition in which a pathogenic microorganism is present and multiplying in another organism.

infection: the invasion of the body by a pathogenic microorganism that reproduces and multiplies, causing disease by local cell injury, the secretion of toxin, or an antigen-antibody reaction in the host.

infectious: capable of causing infection or caused by an infection.

infective: the condition where an organism is capable of spreading infection rapidly to another host.

insect: a member of the class Insecta, which includes all arthropods that have a head free from the thorax, only three pairs of thoracic legs in the adult stage, and a limited number of segments.

insecticide: a chemical agent, sometimes of natural origin, that kills insects.

invertebrate: an animal without a backbone.

irritant: an agent that produces inflammation or irritation.

larva, larvae (pl.): an early stage of development, commonly hatching from the egg stage; the immature stage of organisms which undergo metamorphosis, e.g., caterpillar, maggot, grub, wiggler.

metamorphosis: the series of changes through which an insect passes in its growth from the egg through the larva and pupa to adult; it is termed complete when the pupa is inactive and does not feed; incomplete when there is no pupa or when the pupa is active and feeds.

necropsy: a postmortem examination performed on an animal to confirm or determine the cause of death.

OECD: abbreviation for the Organisation for Economic Co-operation and Development, an international association of 30 countries that accept the principles of representative democracy and a free market economy.

parasite: any organism that lives in, on, or at the expense of another.

pathogen: any microorganism capable of producing disease.

PBO: piperonyl butoxide, a chemical added to insecticide formulations to improve their effectiveness (a synergist).

pesticide: a chemical agent, sometimes of natural origin, used to kill pests, including weeds, insects, arachnids, and rodents; an insecticide is a kind of pesticide.

physiology: the study of the physical and chemical processes involved in the functioning of living organisms and their component parts.

sensitization: an acquired reaction in which specific antibodies develop a response to an antigen. This is deliberately caused in immunization by injecting a disease-causing organism that has been altered in such a way that it is no longer infectious yet remains able to cause the production of antibodies to fight the disease. Allergic reactions are hypersensitive reactions that result from excess sensitization to a foreign protein.

species: an aggregation of individuals alike in appearance and structure, mating freely and producing young that themselves mate freely and bear fertile offspring resembling each other and their parents, including all varieties and races; the primary biological unit, the definition of which is often debated by experts.

synergist: an organ, agent, or substance that augments the activity of another organ, agent, or substance; for pesticides, a chemical that multiplies the effects of the primary toxicant.

systemic: of or pertaining to the whole body, rather than to a localized area or portion of the body.

toxic: description of a chemical that interferes with a physiological process, especially in a way that the process gets progressively worse after administration.

toxicity: the degree to which something is toxic. In addition, a condition that results from exposure to a toxin or to toxic amounts of a substance that does not cause adverse effects in smaller amounts.

toxicology: the scientific study of toxins, their detection, their effects, and the methods of treatment for the conditions they produce.

toxin: a chemical that interferes with a physiological process in a dose-dependent manner.

urticaria: a skin eruption characterized by transient wheals of varying shapes and sizes with well-defined red margins; may be caused by drugs, foods, insect bites, inhalants, emotional stress, exposure to heat or cold, or exercise; also called hives.

vector: an arthropod that transmits microorganisms or parasites to another animal or to a plant, especially pathogens that cause disease. A biological vector is usually an arthropod in which the infecting organism completes part of its life cycle, while a mechanical vector transmits the organism from one host to another but is not essential to the life cycle of the parasite.

venom: a toxic fluid secreted by some snakes, arthropods, and other animals and transmitted by their bites or stings. Venoms can be toxic to some organisms but not to others.

venomous: an animal that possesses venom and is able to transmit this venom by a bite or sting.

vertebrate: any animal possessing a backbone and thus a member of the subphylum Vertebrata, phylum Chordata. The group includes fish, birds, amphibians, reptiles, and mammals.

volatile: a liquid easily vaporized into a gaseous form.

volatility: the degree to which something is volatile.

wheal: an individual lesion of urticaria.

WHO: abbreviation for the World Health Organization.

zoonosis: a disease of animals the pathogen of which is transmissible—sometimes by a vector—to humans from its primary animal host.

REFERENCES

This book was written for the public and for professionals who need answers to problems caused by arthropods that use blood for nutrition or venoms for defense against people. The science, product development, and government regulation surrounding these problems are vast, dynamic, and complicated. There are key general references to the subjects involved, including more than a few that are not mentioned in this list, and there are literally hundreds of thousands of specific references, each of which carves out a few facts gleaned from careful experiments and observation. The authors have relied on many of these references for the material in this book, but we have not attempted to list every one of them. We have also relied heavily on our own experiences in the field as scientists, as practicing entomologists advising people on how to control problems, and on the advice of our colleagues. This reference list is intended to give the reader an introduction to the scientific literature and to some of the general sources that are very useful. Key references are marked with an asterisk (*).

CHAPTERS 1-3

These chapters cover identification, general biology, and the distribution of arthropods that take blood, with only a few comments on the identification of venomous arthropods. They also review the damage caused by the blood-sucking arthropods, either directly to the skin or by transmission of infectious pathogens. The references are grouped together because they tend to provide information for more than one of these biological chapters. The reader who would like to know more about these fascinating arthropods and the complicated ecology of pathogen transmission may want to take a closer look at these books, chapters, and articles.

Beck, W., and Pfister, K. 2004. Occurrence of a house-infesting tropical rat mite (*Ornithonyssus bacoti*) on murides and human beings in Munich: 3 case reports. In German. *Wien Klin Wochenschr.* 116 (Suppl. 4): 65–68.

Becker, N., Zgomba, M., Pedric, D., and Dahl, C. 2003. *Mosquitoes and Their Control.* New York: Springer.

Busvine, J. R. 1977. Dermatoses due to arthropods other than the scabies mite. In Orkin, M., et al., eds., *Scabies and Pediculoses.* Philadelphia: Lippincott.

Caeiro, V. 1999. General review of tick species present in Portugal. *Parasitologia* 41 (Suppl. 1): 11–15.

Curtis, C. F., and Davies, C. R. 2001. Present use of pesticides for vector and allergen control and future requirements. *Med. Vet. Entomol.* 15: 231–235.

Failing, R. M., Lyon, C. B., and McKittrick, J. E. 1972. Pajaroello tick bite: The frightening folklore and the mild disease. *Calif. Med.* 116: 16–19.

Foxx, T. S., and Ewing, S. A. 1969. Morphologic features, behavior, and life history of *Cheyletiella yasguri. Am. J. Vet. Res.* 30: 269–285.

*Goddard, J. 2007. *Physician's Guide to Arthropods of Medical Importance*, 5th ed. Boca Raton, FL: CRC Press.

Keh, B. 1973. Dermatitis in man traced to dog infested with *Cheyletiella yasguri* Smiley (Acari: Cheyletiellidae) in California. *Calif. Vector Views* 20: 77–79.

Kolonin, G.V. 1995. Review of the Ixodid tick fauna (Acari: Ixodidae) of Vietnam. *J. Med. Entomol.* 32: 276–282.

*Mullen, G., and Durden, L. 2002. *Medical and Veterinary Entomology.* Amsterdam: Academic.

Mumcuoglu, Y., and Buchheim, E. 1983. Dermatitis caused by the tropical rat mite (*Ornithonyssus bacoti*) in Switzerland: Case report. *Schweiz Med. Wochenschraf.* 113: 793–795. In German.

Patti, I., and Vacante, V. 1984–1985. Un caso di dermatite umana connessa al infestazione di *Cheyletiella blakei* Smiley (Acarina Cheyletiellidae) su gatto persiano. *Boll. Zool. Agr. Bachic.*, ser. 2, 18: 13–22.

*Pinto, L., Cooper, R., and Kraft, S. K. 2007. *Bed Bug Handbook: The Complete Guide to Bed Bugs and Their Control.* Mechanicsville, MD: Pinto & Associates.

Pratt, H. D. 1975. *Mites of Public Health Importance and Their Control.* U.S. Department of Health, Education, and Welfare, Public Health Service, Center for Disease Control, DHEW Publication No. (CDC) 75-8297.

Punyua, D. K. 1992. A review of the development and survival of ticks in tropical Africa. *Insect Sci. Appl.* 13: 537–544.

Robinson, W. H. 2005. *Handbook of Urban Insects and Arachnids.* Cambridge: Cambridge University Press.

Shelley, E. D., Shelley, W. B., Puja, J. F., and McDonald, S. G. 1984. The diagnostic challenge of nonburrowing mite bites. *J. Am. Med. Assoc.* 251: 2690–2691.

Shepherd, A. J., and Narro, S. P. 1983. The genus *Ornithonyssus* Sambon 1928 in the Ethiopian region: Description of a new species and a redescription of *O. roseinnesi* (Zumpt & Till, 1953) (Acarina, Mesostigmata). *Acarologia* 24: 347–351.
*Swaby, J., and Bowles, D. 2006. *Field Guide to Venomous and Medically Important Invertebrates Affecting Military Operations: Identification, Biology, Symptoms, Treatment.* Version 2.0. Armed Forces Pest Management Board, Silver Spring, MD. www.afpmb.org/pubs/Field_Guide.htm.
Whitaker, J. O., Jr., Walters, B. L., Castor, L. K., Ritz, C. M., and Wilson, N. 2007. *Host and Distribution Lists of Mites (Acari), Parasitic and Phoretic, in the Hair or on the Skin of North American Wild Mammals North of Mexico: Records since 1974.* Harold W. Manter Laboratory of Faculty Publications from the Harold W. Manter Laboratory of Parasitology. digitalcommons.unl.edu/parasitology.facpubs/1.
World Health Organization. 1989. *Geographical Distribution of Arthropod-Borne Diseases and Their Principal Vectors.* WHO/VBC/89.967. Geneva.

CHAPTERS 4-6

These chapters include much of the information usually lumped under the subject of household pest control. The emphasis is again on blood-sucking arthropods, but many of the same techniques apply to venomous species. Armed with the information in the previous chapters, these three chapters tell you what to do next: eliminate sources, kill as many of the pests that remain as possible, and erect barriers to stop the remainder.

Boeke, S. J., Boersma, M. G., Alink, G. M., van Loon, J. J., van Huis, A., Dicke, M., and Rietjens, I. M. 2004. Safety evaluation of neem (*Azadirachta indica*) derived pesticides. *J. Ethnopharmacol.* 94: 25–41.
Canyon, D. V., and Speare, R. 2007. A comparison of botanical and synthetic substances commonly used to prevent head lice (*Pediculus humanus* var. *capitis*) infestation. *Int. J. Dermatol.* 46: 422–426.
*Carson, R. 1962. *Silent Spring.* Boston: Houghton Mifflin.
Goldberg, L. Y., and Margalit, J. 1977. A bacterial spore demonstrating rapid larvicidal activity against *Anopheles sergentii, Uranotaenia unguiculata, Culex univittatus, Aedes aegypti,* and *Culex pipiens. Mosquito News* 37: 355–358.
Jackson, J. K., Horwitz, R. J., and Sweeney, B. W. 2002. Effects of *Bacillus thuringiensis israelensis* on black flies and nontarget macroinvertebrates and fish in a large river. *Trans. Am. Fisheries Soc.* 131: 910–930.
*Metcalf, R. L. 1973. A century of DDT. *J. Agricultural and Food Chem.* 21: 511–519.

Panella, N. A., Dolan, M. C., Karchesy, J. J., Xiong, Y., Peralta-Cruz, J., Khasawneh, M., Montenieri, J. A., and Maupin, G. O. 2005. Use of novel compounds for pest control: Insecticidal and acaricidal activity of essential oil components from heartwood of Alaska yellow cedar. *Pesticide Action Network North America: PAN Pesticide Database.* www.pesticideinfo.org.

Smith, D. 1999. Worldwide trends in DDT levels in human breast milk. *Int. J. Epidemiol.* 28: 179–188.

Sullivan, W. N. 1971. The coupling of science and technology in the early development of the World War II aerosol bomb. *Mil. Med.* (February): 157–158.

CHAPTERS 7-12

These chapters are about personal protection, which includes the measures you have to take when you are either away from home or when the methods described in chapters 4–6 do not provide satisfactory results. The authors of this book edited a technical volume that comprehensively reviewed many aspects of personal protection. The U.S. military maintains two Web sites open to the public that include many excellent nonpublished resources on personal protection.

*Armed Forces Pest Management Board, U.S. Department of Defense. www.afpmb.org.

*Debboun, M., Frances, S. P., and Strickman, D. 2007. *Insect Repellents: Principles, Methods, and Uses.* Boca Raton, FL: CRC Press.

*U.S. Army Center for Health Promotion and Preventive Medicine. chppm-www.apgea.army.mil.

Chapter 7

The references listed for this chapter are old, almost historical. Before the development of a wide variety of chemically based products for personal protection, people gave a lot more thought to the use of clothing for protection. Much of the work has never been repeated, but it is still relevant today.

Elton, N. W., Edinger, O. H., Jr., and Lewis, J. S. 30 October 1944. Subject: Mosquito repellency of 8 standard quartermaster fabrics. Letter to Quartermaster Board, Camp Lee, VA, from 363rd Med Comp Detch (Lab), Camp Ellis, IL.

Linduska, J. P., and Morton, F. A. 2 June 1945. *Laboratory and Field Tests on the Permeability of Fabrics to Biting by Mosquitoes.* Committee on Medical Research of the Office of Scientific Research and Development, OSRD Insect Control Committee Report No. 29, Interim Report No. 0-95.

McWilliams, J. G. 6 March 1968. Resistance of military uniform fabrics to penetration by biting mosquitoes. Letter to Lt. Col. William W. Young, 20th Preventive Medicine Unit (SVC) (Fld) from Armed Forces Pest Control Board.

Sholdt, L. L., Rogers, E. J., Jr., Gerberg, E. J., and Schreck, C. E. 1989. Effectiveness of permethrin-treated military uniform fabric against human body lice. *Military Med.* 154: 90–93.

Travis, B. V., and Morton, F. A. 1946. *Treatment of Clothing for Protection against Mosquitoes.* Proc. 33rd Annual Meeting New Jersey Extermination Assoc.

Whayne, T. F. 1955. Clothing. *Preventive Medicine in World War II*, vol. 3 of Coates, J. B., Jr., and Hoff, E. C., eds., *Personal Health Measures and Immunization.* Office of the Surgeon General, Medical Department, U.S. Army.

Chapter 8

Industry and science explore the use of natural products vigorously because the public has the perception that "natural" is good. The search for effective, safe, natural products as insect repellents has been difficult, to say the least. However you explore the subject, you will find long lists of natural sources for chemicals and extracts and an equally long list of references. Debboun et al., *Insect Repellents: Principles, Methods, and Uses* (listed above), has three chapters on this subject, leading to hundreds of other references. The list below is a sample of the information sources used in this book, which may stimulate the reader to look at the science, as well as the lore, of natural products.

Amer, A., and Mehlhorn, H. 2006. Repellency of forty-one essential oils against *Aedes, Anopheles,* and *Culex* mosquitoes. *Parasitol. Res.* 99: 478–490.

Barnard, D. R. 1999. Repellency of essential oils to mosquitoes (Diptera: Culicidae). *J. Med. Entomol.* 36: 625–629.

Das, N. G., Baruah, I., Talukdar, P. K., and Das, S. C. 2003. Evaluation of botanicals as repellents against mosquitoes. *J. Vector Borne Diseases* 40: 49–53.

Dietrich, G., Dolan, M. C., Peralta-Cruz, J., Schmidt, J., Piesman, J., Eisen, R. J., Karchesy, J. J. 2006. Repellent activity of fractioned compounds from *Chamaecyparis nootkatensis* essential oil against nymphal *Ixodes scapularis* (Acari: Ixodidae). *J. Med. Entomol.* 43: 957–961.

Hach, V., and McDonald, E. C. 1973. Terpenes and terpenoids: IV. Some esters and amides of thujic acid. *Canadian J. Chem.* 51: 3230–3235.

Hadis, M., Lulu, M., Mekonnen, Y., and Asfaw, T. 2003. Field trials on the repellent activity of four plant products against mainly *Mansonia* populations in western Ethiopia. *Phytotherapy Res.* 17: 202–205.

International Fragrance Association. 2007. *Code and Standards*. www.ifraorg.org/GuideLines.asp.

Jaenson, T. G. T., Pålsson, K., and Borg-Karlson, A.-K. 2005. Evaluation of extracts and oils of tick-repellent plants from Sweden. *Med. Vet. Entomol.* 19(4): 345–352.

Jantan, I., and Zaki, Z. M. 1999. Development of environment-friendly insect repellents from the leaf oils of selected Malaysian plants. *ASEAN Review of Biodiversity and Environmental Conservation* (ARBEC), www.arbec.com.my/pdf/art8novdec99.pdf.

Miot, H. A., Batistella, R. F., de Almeida Batista, K., Volpato, D. E. C., Augusto, L. S. T., Madeira, N. G., Haddad, V., Jr., and Miot, L. D. B. 2004. Comparative study of the topical effectiveness of the andiroba oil (*Carapa guianensis*) and DEET 50% as repellent for *Aedes* sp. *Rev. Inst. Med. Trop. S. Paulo* 46: 253–256.

Moore, S. J., Darling, S. T., Sihuincha, M., Padilla, N., and Devine, G. J. 2007. A low-cost repellent for malaria vectors in the Americas: Results of two field trials in Guatemala and Peru. *Malaria J.* 6: 101.

Mumcuoglu, K. Y., Galun, R., Bach, U., Miller, J., and Magdassi, S. 1996. Repellency of essential oils and their components to the human body louse, *Pediculus humanus humanus*. *Entomologia Experimentalis et Applicata* 78: 309–314.

Nishimura, H., and Satoh, A. 2000. Potent mosquito repellents from the leaves of *Eucalyptus* and *Vitex* plants. In Cutler, H. G., and Cutler, S. J., eds., *Biologically Active Natural Products: Agrochemicals*. Boca Raton, FL: CRC Press.

Park, B. S., Choi, W. S., Kim, J. H., Kim, K. H., and Lee, S. E. 2005. Monoterpenes for thyme (*Thymus vulgaris*) as potential mosquito repellents. *J. Am. Mosquito Control Assoc.* 21: 80–83.

Pitasawat, B., Choochote, W., Tuetun, B., Tippawangkosol, P., Kanjanapothi, D., Jitpakdi, A., and Riyong, D. 2003. Repellency of aromatic turmeric *Curcuma aromatica* under laboratory and field conditions. *J. Vector Ecol.* 28(2): 234–240.

Ramanoelina, P. A. R., Viano, J., Bianchini, J. P., and Gaydou, E. M. 1994. Occurrence of various chemotypes in niaouli (*Melaleuca quinquenervia*) essential oils from Madagascar using multivariate statistical analysis. *J. Agric. Food Chem.* 42: 1177–1182.

Satoh, A., Utamura, H., Nakade, T., and Nishimura, H. 1995. Absolute configuration of a new mosquito repellent, (+)-eucamalol and the repellent activity of its epimer. *Biosci. Biotechnol. Biochem.* 59: 1139–1141.

Sukumar, K., Perich, M. J., and Boobar, L. R. 1991. Botanical derivatives in mosquito control: A review. *J. Am. Mosquito Control Assoc.* 7: 210–237.

Thorsell, W., Mikiver, A., and Tunón, H. 2006. Repelling properties of some plant materials on the tick *Ixodes ricinus* L. *Phytomedicine* 13: 132–134.

Trongtokit, Y., Curtis, C. F., and Rongsriyam, Y. 2005. Efficacy of repellent products against caged and free flying *Anopheles stephensi* mosquitoes. *Southeast Asian J. Trop. Med. Pub. Health* 36: 1423–1431.
Tuetun, B., Choochote, W., Kanjanapothi, D., Rattanachanpichai, E., Chaithong, U., Chaiwong, P., Jitpakdi, A., Tippawangkosol, P., Riyong, D., and Pitasawat, B. 2005. Repellent properties of celery, *Apium graveolens* L., compared with commercial repellents, against mosquitoes under laboratory and field conditions. *Trop. Med. & Intl. Hlth.* 10: 1190–1198.
Watanabe, K., Shono, Y., Kakimizu, A., Okada, A., Matsuo, N., Satoh, A., and Nishimurag, H. 1993. New mosquito repellent from *Eucalyptus camaldulensis*. *J. Agric. Food Chem.* 41: 2164–2166.

Chapter 9

The past 50 years have seen the development of new synthetic active ingredients for use in insect repellent products. The inherent purpose of these chemicals creates a tremendous challenge to find the right ones; they must be close to 100% effective for hours and absolutely safe for application directly to skin. The list below emphasizes a few references in addition to or extracted from those reviewed in Debboun et al., *Insect Repellents: Principles, Methods, and Uses*, which includes individual chapters on each of the most effective active ingredients (DEET, DEPA, picaridin, PMD, and IR3535).

Badolo, A., Iboudo-Sanogo, E., Ouédrago, A. P., and Costantini, C. 2004. Evaluation of the sensitivity of *Aedes aegypti* and *Anopheles gambiae* complex mosquitoes to two insect repellents: DEET and KBR 3023. *Trop. Med. & Intl. Hlth.* 9: 330–334.
Buescher, M. D., Rutledge, L. C., and Wirtz, R. A. 1987. Studies on the comparative effectiveness of permethrin and DEET against bloodsucking arthropods. *Pesticide Science* 21: 165–173.
Buescher, M. D., Rutledge, L. C., Wirtz, R. A., Glackin, K. B., and Moussa, M. A. 1982. Laboratory tests of repellents against *Lutzomyia longipalpis* (Diptera: Psychodidae). *J. Med. Entomol.* 19: 176–180.
Christophers, S. R. 1947. Mosquito repellents: Being a report of the work of the mosquito repellent enquiry, Cambridge 1943–[194]5. *J. of Hygiene* 45(2): 176–231.
Kumar, S., Prakash, S., and Rao, K. M. 1995. Comparative activity of three repellents against bedbugs *Cimex hemipterus* (Fabr.). *Indian J. Med. Res.* 102: 20–23.
*McCabe, E. T., Barthel, W. F., Gertler, S. I., and Hall, S. A. 1954. Insect repellents: III. *N,N*-diethyl-amides. *J. Org. Chem.* 19: 493–498. [This was the first publication on DEET.]
Nentwig, G. 2003. Use of repellents as prophylactic agents. *Parasitol. Res.* 90: S40–S48.

Perich, M. J., Strickman, D., Wirtz, R. A., Stockwell, S. A., Glick, J. I., Burge, R., Hunt, G., and Lawyer, P. G. 1995. Field evaluation of four repellents against *Leptoconops americanus* (Diptera: Ceratopogonidae) biting midges. *J. Med. Entomol.* 32: 306–309.

Scheinfeld, N. 2004. Picaridin: A new insect repellent. *J. Drugs Dermatol.* 3: 59–60.

Chapters 10–12

These chapters discuss the details of insect repellent products, including the inactive ingredients that influence the quality of the product, the process of evaluating effectiveness, and the methods used to determine safety. Much of the practical information on these subjects is held in the minds of individuals, a constantly shifting body of regulations, and Web sites of varying quality. As a result, these chapters were more dependent on the experiences of the authors and information we have learned from specialists in industry and government.

Barnard, D. R. 2000. *Repellents and Toxicants for Personal Protection.* World Health Organization Pesticide Evaluation Scheme, Global Collaboration for Development of Pesticides for Public Health, WHO/CDS/WHOPES/GCDPP/2000.5. Geneva.

European Chemicals Bureau. 2004. Methods for the determination of physico-chemical properties, toxicity and ecotoxicity. In *Classification, Packaging and Labeling of Dangerous Substances.* Annex V, Directive 67/548/EEC. Ispra, Italy.

Konan, Y. L., Sylla, M. S., Doannio, J. M., and Traoré, S. 2003. Comparison of the effect of two excipients (karite nut butter and Vaseline) on the efficacy of *Cocos nucifera, Elaeis guineensis* and *Carapa procera* oil-based repellent formulations against mosquitoes biting in Ivory Coast. *Parasite* 10: 181–184.

Mehr, Z. A., Rutledge, L. C., Morales, E. L., Meixsell, V. E., and Korte, D. W. 1985. Laboratory evaluation of controlled-release insect repellent formulations. *J. Med. Entomol.* 21: 665–689.

*U.S. Environmental Protection Agency. 1998. *Registration Eligibility Decision (RED): DEET.* EPA 738-R-98-010. Washington, D.C.

World Health Organization Pesticide Evaluation Scheme. 1996. *Report of the WHO Informal Consultation on the Evaluation and Testing of Insecticides.* CTD/WHOPES/IC/96.1. Geneva

Chapter 13

Venomous arthropods are absolutely fascinating to many people. Some of that interest is a psychological projection of the perceived power of the organisms, but a healthier motivation is an appreciation of the exquisite adaptation of these creatures for subduing their prey and for defending themselves. Most of the information on the biology and pest control of venomous arthropods was concentrated in this chapter. Some of the references listed under chapters 1–3 (Goddard 2007; Mullen and Durden 2002; Robinson 2005; and Swaby and Bowles 2006) include reviews of venomous as well as blood-feeding arthropods.

Alexander, J. O. 1984. *Arthropods and Human Skin*. Berlin: Springer.

Bucherl, W., and Buckley, E. (eds.). 1971. *Venomous Animals and Their Venoms*, vol. 3 of *Venomous Invertebrates*. New York: Academic.

Frazier, C. A., and Brown, F. K. 1980. *Insects and Allergy and What to Do about Them*. Norman: University of Oklahoma Press.

Grismado, C. J., and Goloboff, P. A. 2006. Descripcíon del macho de *Missulena tussulena* Goboloff 1994 (Araneae, Mygalomorphae, Actinopodidae). *Rev. Mus. Argentino Cienc. Nat.* 8: 101–104.

Keegan, H. L. 1969. Some medical problems from direct injury by arthropods. *Int. Pathol.* 10: 35.

——. 1980. *Scorpions of Medical Importance*. Jackson: University Press of Mississippi.

Kjellesvig-Waering, E. N. 1972. *Brontoscorpio anglicus*: A gigantic lower Paleozoic scorpion from central England. *J. Paleontology* 46: 39–42.

*Leeming, J. 2003. *Scorpions of Southern Africa*. Cape Town, South Africa: Struik.

Maschwitz, U. W., and Kloft, W. 1971. Morphology and function of the venom apparatus of insects: Bees, wasps, ants, and caterpillars. In Bucherl, W., and E. Buckley, eds., *Venomous Animals and Their Venoms*, vol. 3 of *Venomous Invertebrates*. New York: Academic.

Nelder, M. P., Paysen, E. S., Zungoli, P. A., and Benson, E. P. 2006. Emergence of the introduced ant *Pachycondyla chinensis* (Formicidae: Ponerinae) as a public health threat in the southeastern United States. *J. Med. Entomol.* 43: 1094–1098.

Southcott, R. 1987. Moths and butterflies. In Covacevich, J., Davie, P., and Pearn, J., eds., *Toxic Plants & Animals: A Guide for Australia*. South Brisbane, Queensland, Australia: Queensland Museum.

Strickman, D., Sithiprasasna, R., and Southard, D. 1997. Bionomics of the spider, *Crossopriza lyoni* (Araneae, Pholcidae), a predator of dengue vectors in Thailand. *J. Arachnology* 25: 194–201.

Sutherland, S. K. 1983. Spider bites in Australia: There are still some mysteries. *Med. J. Aust.* December 10(24): 597.

*Vetter, R. S., and Isbister, G. K. 2008. Medical aspects of spider bites. *Ann. Rev. Entomol.* 53: 409–429.

Visscher, P. K., Vetter, R. S., and Camazine, S. 1996. Removing bee stings: Speed matters, method doesn't. *Lancet* 348: 301–302.

INDEX